International Development Cooperation of Japan and South Korea

Huck-ju Kwon · Tatsufumi Yamagata ·
Eunju Kim · Hisahiro Kondoh
Editors

International Development Cooperation of Japan and South Korea

New Strategies for an Uncertain World

palgrave
macmillan

Editors
Huck-ju Kwon
Seoul National University
Seoul, Korea (Republic of)

Eunju Kim
Hansung University
Seoul, Korea (Republic of)

Tatsufumi Yamagata
Ritsumeikan Asia Pacific University
Beppu, Oita, Japan

Hisahiro Kondoh
Saitama University
Saitama, Japan

ISBN 978-981-16-4600-3 ISBN 978-981-16-4601-0 (eBook)
https://doi.org/10.1007/978-981-16-4601-0

This Palgrave Macmillan imprint is published by the registered company Springer Nature Singapore Pte Ltd.
The registered company address is: 152 Beach Road, #21-01/04 Gateway East, Singapore 189721, Singapore

Acknowledgments

This book is an outcome of collaboration between the Japan Society for International Development (JASID) and the Korean Association of International Development Cooperation (KAIDEC). The JASID and KAIDEC have worked together closely on both academic and policy discussions. Members of the two associations have participated in annual conferences hosted by their counterparts. During the period of the book project, the relationship between two Northeast Asian neighbors was unusually cold, granting that there have always been ups-and-downs in the relationship. Nevertheless, the JASID and KAIDEC maintained its close working relationship. This book epitomizes the underlying cooperation beneath the international relations. Two members of the editorial team, Huck-ju Kwon and Tatsufumi Yamagata, were the Presidents of respective associations, KAIDEC and JASID, which provided strategic advantage in assembling scholars and experts from two countries. There have been international exchanges through which Japanese and Korean authors could meet regularly, with some meetings organized by the United Nations Economic Commission of Asia and the Pacific (UN ESCAP) Subregional Office for East and North-East Asia. The Asia Development Institute at Seoul National University provided a research grant for the Korean side of the project and hosted workshops for contributors in 2020 and 2021. It would not have been possible to publish this book without such previous experiences and collaborations. Editors are profoundly grateful for these kind arrangements.

Contents

Notes on Contributors

Jiyoung Hong is a Senior Research Fellow at Research Institute of North Korea and Northeast Asia Development of Korea Eximbank. She received her Ph.D. in political science at Seoul National University. Her research interests include international development cooperation, North Korea development cooperation, humanitarian assistance, and transnational/non-governmental actors, with a focus on international norms and cooperation mechanisms.

Hyomin Jung is a Post-Doctoral Research Fellow at the Graduate School of Asian and African Area Studies, Kyoto University. She is also a Lecturer at the Konan Women's University. Her research interests focus on international development policies and donors.

Sung Gyu Kim obtained Dr. Phil. in sociology from Free Uni. (FU) Berlin and was a researcher of Otto-Suhr-Institute of FU Berlin until 2007. In Korea he was an expert committeeman of Economic and Social Development Commission of Korea and Senior Research Fellow of KOICA (Korea International Cooperation Agency) and Seoul National University Asia Center (SNUAC). He works as a director at Center for Sustainable Development, Asiatic Research Institute, Korea University. His main research areas include ODA policy of South Korea and international development cooperation.

Taekyoon Kim is a Professor of International Development at the Graduate School of International Studies, Seoul National University. His focus areas are international development, global social governance, and international political sociology. He has published articles in multiple prestigious journals and books on the topics of reflection on development theories and accountability mechanisms in Asia.

Bo Kyung Kim is a Lecturer and Post-Doctoral Research Fellow at the Graduate School of International Studies, Seoul National University. Her publications mainly focus on her major areas of interest, including international development, foreign aid, global governance, and diverse forms of accountability mechanisms in development cooperation.

Eun Mee Kim is the 17th President of Ewha Womans University and a Professor at the Graduate School of International Studies at Ewha Womans University. She is also the Director of Ewha Global Health Institute for Girls and Women. She was appointed by the UN Secretary General in December 2016 to serve as one of the fifteen Independent Group of Scientists to work on the Global Sustainable Development Report 2019, which was presented to the UN Secretary General and heads of member states at the UN General Assembly in September 2019.

Eunju Kim is an Assistant Professor at the College of Social Science, Hansung University, South Korea. She obtained her Ph.D. in public policy from the Graduate School of Public Administration, Seoul National University. Previously, she was a Research Fellow at the Korea Institute of Public Administration and a visiting fellow at the Centre for the Study of Poverty and Social Justice, University of Bristol, UK. Her research interests include international development policy and global governance, policy evaluation, poverty measurement, and social protection in developing countries.

Hisahiro Kondoh is a Professor of the Graduate School of Humanities and Social Sciences, Saitama University. He was awarded his master's degree from the School of Development Studies, University of East Anglia, UK, and his doctoral degree from the Institute for Politics and International Studies, University of Leeds, UK. He has expertise in the analysis of political economy of development and emerging donor issues. Besides his academic career, he has working experience of the civil service in local government and professional development consultancy.

Huck-ju Kwon is a Professor at the Graduate School of Public Administration and the Asia Development Institute, Seoul National University. He is the former President of the Korean Association of International Development and Cooperation (2019–2020) and is a Fulbright-Democracy Visiting Scholar at Harvard Kennedy School. His publications include "Normative Rationale for a Relationship between Political Power and Bureaucracy: The Perspective of Democracy and Republicanism" *(Korean Journal of Public Administration*, 2020), "Bring back institution capability in fragile situations: Implications for global social policy" (*Global Social Policy*, 2018), and *The Korean Government and Public Policy in a Development Nexus* (Springer, 2017).

Izumi Ohno is a Professor of the National Graduate Institute for Policy Studies (GRIPS). She specializes in international development policy. Prior to joining GRIPS in 2002, she worked at the Japan International Cooperation Agency (JICA), the World Bank, and the Japan Bank for International Cooperation (JBIC). She also served as Director of the JICA Ogata Sadako Research Institute for Peace and Development during 2018–2020.

Kyung Ryul Park is an Assistant Professor in the Graduate School of Science and Technology Policy, and an Adjunct Professor in the College of Business at Korea Advanced Institute of Science and Technology (KAIST). His research is primarily concerned with questions of technology, governance and global development. He received his Ph.D. in Management—Information Systems and Innovation from London School of Economics and Political Science (LSE).

Jin Sato is a Professor at the Institute of Advanced Studies on Asia of the University of Tokyo. He served as a visiting professor at the School of Public and International Affairs in Princeton University from 2017 to 2020. He is currently the President of Japan Society for International Development and Chief-Editor of the International Journal of Asian Studies.

Jisun Song is Assistant Professor at the Institute of Foreign Affairs and National Security of the Korea National Diplomatic Academy. Previously, she worked as Research Professor at the Institute for Development and Human Security of Ewha Womans University. She received her Ph.D. in International Studies from Ewha Womans University. Her research areas and publications center around international development cooperation, aid effectiveness, foreign aid modalities, global health security, and gender and development.

Motoki Takahashi is a Professor at the Graduate School of Asian and African Area Studies, Kyoto University. He is also the Director of the Center for African Area Studies, Kyoto University and a Professor Emeritus, Kobe University. He has published a number of academic books and articles on African development and Japan's Official Development Assistance in Japanese and other languages.

Akio Takayanagi is a Professor at the Department of Global and Intercultural Studies at Ferris University. He is also a Policy Advisor of Japan NGO Center for International Cooperation (JANIC). His research has focused on the roles of civil society organizations (CSOs) in international development and enabling environment for CSOs. In 1993–2002, he was Assistant/Associate Professor of the Department of International Relations, Faculty of Foreign Studies at Kitakyushu University.

Shinichi Takeuchi is the Director of the African Studies Center at Tokyo University of Foreign Studies. He is also a Senior Research Fellow at the Institute of Developing Economies—JETRO. He holds a Ph.D. from the University of Tokyo and his research focuses mainly on African politics and international relations.

Sayoko Uesu is Research Associate of the National Graduate Institute for Policy Studies (GRIPS). She specializes in political economy and security of sub-Saharan Africa, especially francophone countries in West Africa. She also worked at the Japan International Cooperation Agency (JICA) and Japan's Ministry of Foreign Affairs (MoFA), and in several postings in France as advisor for African issues.

Tatsufumi Yamagata is a Professor of the Ritsumeikan Asia Pacific University. He is a former President of the Japan Society for International Development, 2017–2020. He was a Research Fellow of the Institute of Developing Economies, 1988–2018. He earned Ph. D. in economics from the University of Rochester.

List of Figures

Chapter 3

Chapter 4

Chapter 6

Chapter 7

Chapter 8

Chapter 9

Chapter 10

Chapter 11

Chapter 12

List of Tables

Chapter 4

Chapter 5

Chapter 6

Chapter 7

Chapter 8

Chapter 9

Chapter 10

Chapter 11

Chapter 12

Introduction

Aims of the Book

East Asia has increasingly played a significant role in international development cooperation. While China has embarked on an ambitious Belt-and-Roads Initiative, which includes elements of development assistance as well as those relating to outright commercial loans (Yu, 2017), Japan and Korea, as East Asia's only two members of the influential Development Assistant Committee of the Organization for Economic Cooperation and Development (OECD/DAC), have shared responsibility for supporting developing countries toward sustainable development with official development assistance (ODA) following international norms and rules.

Japan has provided a large amount of foreign aid to developing countries for the last several decades, although the size of its ODA has declined in recent years. In 2019, Japan provided USD 15.5 billion in ODA, which was the fourth largest in size among OECD/DAC member countries. Despite such a contribution to international development cooperation, Japanese ODA has been criticized for its lack of clear policy goals, fragmented administrative structure, and weak involvement of civil society organizations (Lancaster, 2007a). Other commentators point out that commercial consideration has been a main concern for Japanese foreign aid (Hulme, 2016; Söderberg, 1996). In response to those criticisms, Japan put forward the notion of human security as an ideal for international development cooperation and reorganized its administrative structure for effective and efficient implementation in the

2000s. In 2015, however, the Japanese government announced a kind of volte face in ODA policy. In its new Official Development Assistance Charter, the Abe government explicitly established national interests as the main goal of foreign aid.

In contrast to Japan's relatively long experience as a donor country, Korea only joined the OECD/DAC in 2010. Nevertheless, it was considered a remarkable transformation from a war-torn poor society to an affluent industrialized country with a sense of responsibility as a member of the international community. Policy makers in many developing countries see the Korean experience of development as relevant in devising their development strategies (Kohli, 2004). In terms of size, Korean ODA was 15th among the OECD/DAC members, equivalent to USD 2.5 billion in 2019. This corresponded to 0.15% of Korea's gross national income (GNI). Although this is less than what Korea promised the OECD/DAC when it joined, the size of ODA has steadily increased over recent years. With a decade of experience since joining the OECD/DAC in 2010, Korea has tried to move beyond its rather limited role as an emerging donor. Despite its impressive social transformation and increasing role in international development, there have also been various criticisms directed at Korean ODA. While some commentators argue that Korean ODA does not have a clear policy rationale in terms of goals, strategies, and instruments (Jung et al., 2018), others criticize that it has been too influenced by economic considerations (Hulme, 2016; Kim and Oh, 2012). Although the Korean government is preparing its third Mid-Term Strategic Plan for International Development Cooperation, it is expected to focus on achieving the Sustainable Development Goals (SDGs) and enhancing the aid effectiveness of Korean ODA as well as putting forward national interest.

On the global stage, these two East Asian countries are expected to play bigger roles in the global political and economic competition between superpowers. The US has shifted its focus to national interests rather than global economic stability, protecting domestic industries and workers. Although newly elected President Biden is trying to change policy directions, the US government under President Trump reduced the military budget for strategic interests abroad, creating friction with US allies. Meanwhile, China has tried to assert its ambition as a rising superpower. The Chinese expansionist geopolitical strategy in the South China Sea has created military tension in the region (Morton, 2016). Its Belt-and-Roads Initiative has been implemented in Central Asia, South

Asia, and Africa, although it has produced less than substantial economic outcomes for developing countries in these regions. The trade disputes between the US and China add further tension to the already tense global political and economic situation. Amid such tensions between the two superpowers, the world's middle powers such as Germany, France, and Britain need to play a significant role to balance the boat. In East Asia it is obviously Japan and Korea that are expected to assume roles as middle-power countries in the uncertain world. Japan, the world's third largest economy, and Korea, the tenth largest trading economy, have no choice other than taking their fair share of responsibility for creating a stable global political order and economic environment.

It is true that the global coronavirus pandemic has made it more difficult for the world to create a stable and prosperous global order. At the same time, however, the coronavirus crisis has shown that Japan and Korea should play unique roles to support developing countries in coping with the coronavirus pandemic and to continue efforts to reduce poverty in the coming global economic recession. This is not only because Japan and Korea have come through the coronavirus pandemic relatively unscathed compared to other developed countries such as the US, UK, and Italy, but also because Japan and Korea have shown clear capabilities to support developing countries in combating the epidemic with antiviral medicine and other medical equipment (Yamagata, 2020). It is also true that Japan and South Korea were not able to play a leading role in overcoming the COVID-19 pandemic: Japan and Korea could not develop and produce COVID-19 vaccines, while China and Russia provided vaccines to developing countries.

These observations lead us to ask three important questions about Japanese and Korea ODA policy.

1. What are the underlying dynamics of policy making in Japanese and Korean ODA?
2. Will national interests be the goal of international development cooperation, or will creating the global public good be the main objective of these countries' ODA policies?
3. Do Japan and Korea have new ODA strategies for an uncertain world?

These questions are related to each other since the answers to the first question will lead us to those of the second and third question. In particular, the first question is about the evolution of ODA policy, which includes both intentional policy changes and unintended drifts. Although Japanese and Korean ODA has been criticized for strong economic considerations, this is only one aspect of the policy rationale for ODA. Other considerations such as creating the global public good and self-regarding values can be important factors. They are not always mutually exclusive and can be present in the minds of policy makers and aid practitioners at the same time as policy rationale. Why, then, are some ideas predominant in ODA policy while other considerations remain in the margins?

Although Japan and Korea have demonstrated certain characteristics in their international development policies (Kim and Oh, 2012; Lancaster, 2007b), policy makers and experts have contemplated new policy strategies in order to achieve effective outcomes in international development, including the SDGs. It is also true that there have been reforms of the international development systems in both countries. Japan has tried to exert greater influence over global affairs as a middle-power country amid increasing confrontation between the US and China. Meanwhile, Korea wants to upgrade its standing in the international development community after spending a decade as an OECD/DAC member. Do these changes amount to a paradigm shift, policy changes in the existing paradigm, or a simple shift of gear in the same set of systems (Hall, 1993)?

According to Hall (1993), policy changes can be divided into three levels of changes. First, paradigm changes refer to changes that happen in the form of overarching goals and a theoretical framework. Since policy goals guide policy orientations in the fields concerned in terms of policy instruments to achieve those goals, there must be a coherent theoretical framework such as Keynesian economics or neoliberal economic theory. In international development cooperation, different kinds of theoretical paradigms such as theories of realism and ideas of international society can guide policy directions. Different theoretical paradigms may define conceptions of national interests and the global public good differently, and policy orientations can accordingly be different. Secondly, changes can happen at the level of policy instruments. For instance, foreign aid can be delivered by bilateral agencies such as the JICA and KOICA or by multilateral organizations such as the UN agencies and development banks. Governments can choose different options given their ODA policy paradigm. The third level of changes involves parameter changes such as changes in ODA expenditure year by year. Although these changes are also important, they can take place without changes in policy instruments and overarching goals.

Given that Hall's model offers a very useful framework to analyze policy changes, we do not assume that policy practices in the real world conform to such a rational policy model in which policy programs are designed and arranged according to a policy paradigm. Policy makers often decide to do whatever they can in order to deal with immediate challenges and demands. Once they muddle through those immediate tasks, they begin to consider the policy rationale which might underlie those programs they have implemented. This book establishes a working hypothesis that Japanese and Korean ODA began with imperatives to deal with immediate policy challenges and that these countries tried to conceive a policy paradigm afterwards in order to comply with global norms and national demands.

It is also important to note that policy changes can happen through policy learning. Political decisions and domestic social demand as well as external pressure can affect policy changes, but policy learning is also an important factor to influence policy changes. In particular, policy learning has been important in Japan and Korea since they had learned from donor countries in Europe and North America in the field of international development cooperation. As the chapters in this book will show, such a learning process has also taken the form of self-reflections and policy transfers as Japan and Korea have tried to construct a policy rationale by looking at the shortcomings as well as the strengths of their experiences.

Following Peter Hall's conception of policy paradigm and change (1993), the book will take on these questions in the following scope of research: policy rationale, institutional development, and policy alternatives and choices. The book will also identify new challenges for international development and try to suggest new strategies and future directions that Japan, and Korea should take in the face of an uncertain world.

The Structure of the Book: Comparative Studies but Separate Analyses

Despite the emergence of two East Asian countries, Japan and Korea, as significant players in the international development community, comparative studies that deal with public policies for international development in Japan and Korea are few and far between, with a few exceptions (Kato et al., 2016; Stallings, 2017). There is a significant chasm in the literature related to international development. One of the main reasons for such a gap is that there is so much difference in ODA between Japan and Korea in terms of historic experiences, size, and global strategies, despite certain similarities in institutions and policies for international development (Stallings, 2017). For instance, the size of Japanese ODA was six times bigger than that of Korea in 2019. Japan has a long experience with international development cooperation, and as one of founding members of the OECD/DAC in the 1960s, Japan carries significant weight in the global paradigm of international development policy. In contrast, Korea is still a relatively small donor with a 10-year membership in the OECD/DAC. Such differences make one-to-one comparisons of the international development policies in these countries difficult. At the same time, there would be a great deal of advantage in learning about international development cooperation policies by comparing these two countries.

To overcome such difficulties, this book will compare Japan and Korea according to the same themes while addressing them in separate chapters. For instance, contributing scholars and experts from Japan and Korea will attempt to analyze each country's public policies and institutions in their own contexts, but will try to answer the same set of questions related to both countries' international development: the overarching question will be whether these two countries have changed their policies for international development over the last decade and what sort of international development policy they should pursue in the face of the uncertain world.

In this book we will approach ODA policy and practice in Japan and Korea through the perspective of the ODA ecosystem, in which government ministries and policy makers are only one component of the whole system. It also has different constituents, such as platforms, agencies, and individuals, that coexist (Hanna et al., 2011). Although international development cooperation is largely dependent on public agencies and resources, there are other actors and resources, for instance private firms, civil society organizations, academic institutes and experts, philanthropic foundations, and volunteers as individual agent. These actors and institutions are independent but work closely together. Even though policy makers in the government decide a certain policy goal and adopt various tools, effective outcomes depend on whether other actors and institutions have successfully done their parts. Policy experts need to design the required programs and projects to support developing countries, and ODA practitioners, both private and public, should be able to deliver those policy programs and projects in the field (Lipsky, 1980). If these actors and institutions fail to play their roles in the process of ODA implementation, policies with good intentions will not achieve their goals. The perspective of the ODA ecosystem is particularly relevant in Japan and Korea where civil society organizations, private firms, and academic institutions are still weak in the field of international development cooperation. The ecosystem perspective will also enable us to see that there are different ideas and goals for international development cooperation that exist at the same time. Sometimes different actors and institutions pursue contradictory ideas and goals, while most of the time they try to collaborate with each other.

Based on the idea of the policy paradigm and ecosystem, the book will be structured in three parts. After Introduction, Part 1 is on policy rationales and institutional development, and it includes two pairs of chapters. The first pair of chapters discusses the policy rationale of ODA in Japan and Korea. In Chapter 1, Shinichi Takeuchi examines policy concepts and the normative rationale in Japan's foreign aid with a focus on human security together with Japan's regional initiatives such as the Tokyo International Conference on African Development (TICAD) and Free and Open Indo-Pacific. Chapter 2 by Huck-ju Kwon reflects on the normative rationale for Korean ODA policy, shedding light on duty, self-regards, and obligation. The second pair of chapters in Part 1 examines historical evolution in Japanese and Korea ODA. In Chapter 3, Jin Sato explains the development of Japanese ODA through pivotal moments from the 1950s to the 2010s. Eun Mee Kim and Ji-sun Song examine the evolution of Korea's ODA as a foreign policy instrument in Chapter 4.

In Part 2, which discusses the harmonization of public actions and private initiatives, there are also two pairs of chapters. The first pair of chapters illustrates how these two countries have tried to incorporate different policy orientations into a consistent framework. Chapter 5 by Jung Hyomin and Motoki Takahashi examines Japanese efforts to incorporate two different strands of aid in the 1990s and 2000s: the international mainstream of aid practices and the Asian model of aid–trade–investment. Eunju Kim's Chapter 6 highlights Korea's efforts to strike a balance between universal values and economic interests in development cooperation. The second pair discusses emerging trends in ODA with such ambivalent characteristics. Izumi Ohno and Sayoko Uesu, in Chapter 7, examine new partnerships with the private sector in Japanese international development cooperation, while Kyung Ryul Park, in Chapter 8, investigates how science, technology, and innovation have emerged in Korea's ODA policy.

In Part 3 on inclusive partnerships and new challenges for international development, the first pair of chapters examines changes in partnerships between governments and civil society organizations in international development. Chapter 9 by Akio Takayanagi examines the relationship between the Japanese government and civil society organizations as partners and critics. Chapter 10, written by Sung Gyu Kim and Jiyoung Hong, analyzes changes in the roles of Korean development civil society organizations (CSOs) in the historical path of Korean international development cooperation since the 1990s. The second pair of chapters explores possible options for the implementation of sustainable development with a focus on international development cooperation. Tatsufumi Yamagata, in Chapter 11, examines government and private sector approaches to the Sustainable Development Goals and highlights decoupling tendencies in the relationship between the SDGs and international development. Taekyoon Kim and Bo Kyung Kim, in Chapter 12, examine government policies on nationalizing the SDGs, shedding light on the relationship between the government and the private sector, and explore ways to reinforce the emerging institutional framework in which public and private actors have begun to collaborate toward achieving the SDGs.

In the conclusion, it tries to answer the questions posed in this introduction by summarizing the chapters in the book. Based on such a discussion, it also speculates on Japan's and Korea's new strategies of international development cooperation for an uncertain world. It will discuss such new strategies in the context of a liberal international order under pressure from two contradictory directions: a series of unilateral actions by the US and China's quite assertive policy orientations toward global order. In particular, it will explore roles in international development cooperation which Japan and Korea can play as middle-power countries.

Huck-ju Kwon
Tatsufumi Yamagata
Eunju Kim
Hisahiro Kondoh

References

Hall, Peter A. (1993). Policy paradigms, social learning, and the state: The case of economic policymaking in Britain, *Comparative Politics, 25*(3), 275–96.

Hanna, Richard, Rohm, Andrew, and Crittenden, Victoria L. (2011). We're all connected: The power of the social media ecosystem, *Business Horizons, 54*(3), 265–73.

Hulme, David (2016). *Should rich nations help the poor?*. Cambridge University Press.

Jung, Changkuk, Cho, Wonbin, and Hwang, Wonjae (2018). Does official development assistance promote foreign policy cooperation from its recipients? The case of South Korea, *Pacific Focus, 33*(1), 83–110.

Kato, Hiroshi, Page, John, and Shimomura, Yasutami (eds.) (2016). *Japan's Development Assistance: Foreign Aid and the Post-2015 Agenda*. Palgrave, 369.

Kim, Eun Mee and Oh, Jinhwan (2012). Determinants of foreign aid: The case of South Korea, *Journal of East Asian Studies, 12*(2), 251–274.

Kohli, Atul (2004). *State-Directed Development: Political Power and Industrialization in the Global Periphery*. Cambridge University Press.

Lancaster, Carol (2007a). Japan: The rise and decline of an "Aid Superpower", in Carol Lancaster (ed.), *Foreign Aid: Diplomacy, Development, Domestic Politics*. Chicago University Press, 110–142.

——— (2007b). *Foreign aid: Diplomacy, development, domestic politics*. University of Chicago Press.

Lipsky, Michael (1980). *Street-level bureaucracy: Dilemmas of the individual in public services*. Publications of Russell Sage Foundation; Russell Sage Foundation, xviii, 244p.

Morton, Katherine (2016). China's ambition in the South China Sea: Is a legitimate maritime order possible? *International Affairs, 92*(4), 909–940.

Söderberg, M., Staff Corporate Author. (1996). *The business of Japanese foreign aid: five case studies from Asia*, ed. Marie Söderberg, Routledge.

Stallings, Barbara (2017). *Promoting development: The political economy of East Asian foreign aid*. Springer Singapore.

Yamagata, Tatsufumi (2020). Japan's international cooperation to cope with COVID-19, *International Development and Cooperation Review, 12*(3), 17–22.

Yu, Hong (2017). Motivation behind China's 'One Belt, One Road' initiatives and establishment of the Asian infrastructure investment bank, *Journal of Contemporary China, 26*(105), 353–368.

PART I

Policy Rationale and Evolution of the Development Institutions

CHAPTER 1

Policy Concepts and Normative Rationales in Japan's Foreign Aid: Human Security, TICAD, and Free and Open Indo-Pacific

Shinichi Takeuchi

1 Introduction

Countries have various motivations for providing foreign aid. Hume (2016, 12–17) groups them into four categories: moral obligations, moral responsibility, common interests among human beings, and short-term political and commercial advantages. When providing official development assistance (ODA), governments consider all of these points. In addition, governments often need to express their philosophy and principles on foreign aid. In both diplomatic relations and communications with national taxpayers, it is often essential for governments to explain why they are conducting such altruistic acts. Given this context, governments request and establish policy concepts that connect various motivations

S. Takeuchi (✉)
African Studies Center, Tokyo University of Foreign Studies, Fuchu, Japan
e-mail: shinichi_takeuchi@tufs.ac.jp

H.-j. Kwon et al. (eds.), *International Development Cooperation of Japan and South Korea*,
https://doi.org/10.1007/978-981-16-4601-0_1

and daily practices with normative rationales and philosophical positions on aid.

Since the 1990s, Japan has made serious efforts to clarify the philosophical basis of its foreign aid policy. Two ODA charters, drafted in 1992 and 2003 (Government of Japan, 1992, 2003), as well as a charter for development cooperation, drafted in 2015 (Government of Japan, 2015), were important consequences of these efforts. In addition, Japan has actively advocated for several ODA-related policy concepts. The most important concepts concern "human security," the Tokyo International Conference on African Development (TICAD), and Free and Open Indo-Pacific (FOIP). All three policy concepts were discovered, established, and strongly promoted by the Japanese government during this period. Although their content, targets, and contexts differ, all are closely connected with Japan's philosophy of foreign aid and diplomacy in general. What is the background of Japan's constant efforts to clarify its philosophical position on foreign aid? How should its efforts be evaluated? This study attempts to answer these questions.

The first section explores the history of Japan's ODA, which began in the 1950s, and investigates the international environment that influences the country's philosophy and foreign-aid practices. The second section examines the above-mentioned three policy concepts. Interestingly, the meanings and implications of all the three concepts have been transformed over time. Through an analysis of these conceptual changes, this section sheds light on Japan's position in the international community. The third section discusses the outcomes and challenges of Japan's foreign-policy in searching for a normative rationale for aid.

2 Searching for a Philosophical Position on Aid

2.1 *Prior to the 1980s: From Post-War Recovery to Becoming a Major Donor*

It is well known that the origins of Japan's foreign aid were closely related to World War II reparations. The conclusion of the Treaty of Peace with Japan in 1951 imposed an obligation to make reparations for the Allied Powers. While the Treaty stipulated that reparations should be made through the "services of the Japanese people," the Allied Powers were also paid through the provision of Japanese products.[1] A number of infrastructure projects, including the Baluchaung hydropower plant in Myanmar and the development of the Brantas River basin in Indonesia, began as reparations, carried out by Japanese engineers using exported materials produced in Japan. These infrastructure projects, which began as reparations, were later transformed into Japan's representative ODA projects (Shimomura, 2020, 46).

At this stage, Japan referred to its foreign aid as "economic cooperation" (keizai kyoryoku); it consisted mainly of export and import credits, technical cooperation, and the Yen loans. Japan also signed agreements on reparations and economic cooperation with other Asian countries. In fact, Japan's wartime reparations and economic cooperation with Asian countries were implemented in tandem (Suehiro, 2005, 231). Both reparations and economic cooperation promoted exports of Japanese products, providing huge benefits for Japan's private sector and playing a critical role in the country's post-war recovery (Arase, 1995). During this early post-war period, it was evident that Japan was designing and building a system of economic cooperation to support its own economic recovery, even if the Japanese people had varied motives for helping Asian countries.[2]

During the 1960s, Japan's economic cooperation was severely criticized for its commercialism. This was the period when Japan joined the Development Assistance Committee (DAC) and became a member of the Organization for Economic Cooperation and Development (OECD).[3] In contrast to other donor countries, Japan was blamed for using aid to benefit itself (by promoting its own exports), rather than recipient

[1] Treaty of Peace with Japan, Article 14, (a) 1. Although the U.S. had initially made strict demands for reparations immediately after World War II, its policies toward Japan softened as the Cold War intensified. As a result, Japan's obligation to make reparations was considerably reduced and the wording, "payment through the 'services of the Japanese people'" was officially introduced into the Peace Treaty. Following the rapid recovery of its industries, Japan preferred to pay other countries by contributing products (Shimomura, 2020, 40–46).

[2] Sincere repentance for wrongdoings and exploitation in Asia during the pre-war and wartime periods has often been considered an important motivation for Japanese people who later worked in and for Asian countries (Araki, 2020, 21; Shimomura, 2020, 33–34).

[3] In 1960, Japan was invited to join the Development Assistance Group (DAG), which turned into DAC in 1961 when the OECD was officially founded. Japan became a member of the OECD in 1964.

countries (White, 1964).[4] Commentators also argued that, even when Japanese ODA brought economic growth to recipient countries, their economies were controlled from Tokyo by Japanese multinational corporations (Hedberg, 1970, 257). During a visit by Prime Minister Tanaka in 1974, anti-Japan movements in Southeast Asian countries revealed clear regional dissatisfaction and distrust. Japan's rapid economic growth was increasingly perceived as a threat by other countries, while its economic cooperation was regarded as a clever device, implemented for its own benefit.

In this context, external pressure was a critical factor shaping Japan's ODA policy. Facing fierce criticism during the 1970s and 1980s, Japan made serious efforts to improve the quantity and quality of its foreign aid by making commitments to significantly increase ODA and to reduce tied loans. Between 1978 and 1993, Japan published five consecutive mid-term plans to increase ODA, expanding its net disbursement more than five times. To enhance quality, following a joint communiqué with the U.S. in January 1978, in which Japan committed to reducing its tied loans, the country rapidly increased its untied loans, achieving the highest percentage of untied loans among all major-donor countries during the 1990s. As a consequence of its efforts to balance its regional allocation of foreign aid, Japan's ODA for African countries increased markedly during the 1970s. External pressure, particularly from the U.S., was a critical factor behind these changes. When the U.S.-Japan trade conflict intensified sharply during the 1980s, the U.S. impatiently demanded that Japan increase its ODA to compensate for the trade surplus (Shimomura, 2020, 149–154).

The Ministry of Foreign Affairs (MOFA) (1981) made the first attempt to explain publicly the Japanese government's rationale for providing ODA. To explain why Japan had to extend foreign aid, this pamphlet distinguished between "internationally accepted philosophies for foreign aid" and "Japan's unique reasons"[5] for providing it. While recognizing the importance of mainstream arguments for contributing foreign aid, expressed in internationally renowned reports, including Pearson (1969)

[4] After publication, this critical book was immediately translated into Japanese by the Economic Cooperation Bureau of the MOFA. This showed that the government was quite sensitive to criticism from abroad.

[5] As there is no English version of MOFA (1981), the quoted passage was translated by the author.

and the "Brandt Report" (Independent Commission on International Development Issues, 1980), this pamphlet stressed that "general philosophy will not change the reality unless it is perceived as something directly affecting our own interests" (MOFA, 1981, 75). In other words, it argued that internationally accepted philosophical justifications for providing foreign aid, such as humanitarian and moral considerations and interdependence, had to be tailored to the context of each country. It argued that Japan needed to extend foreign aid as the "cost" of ensuring its own security. As a peace-loving country and a great economic power, Japan was critically in need of a stable international order, to which increasing both the volume and quality of its ODA would make a significant contribution. In addition, providing ODA would redress Japan's vulnerability, caused by its excessive external dependence. Thus, ODA would improve Japan's image, as well as strengthening developing economies (MOFA, 1981, 76–82).

Yasutomo (1986) criticized these arguments harshly, calling them both ambiguous and confusing because they combined humanitarian and moral considerations with interdependence and strategic purposes. However, the document may be easier to understand if one considers the strong pressure that Japan was experiencing from the international community, particularly from the U.S. The reasoning used in the document may have been exactly the same as the arguments used by the MOFA to persuade the Ministry of Finance to increase ODA. Concretely, the "costs" of the international order stood for the expenditure needed to alleviate U.S. pressure.

2.2 *From the 1990s to the Mid-2000s: A New Era*

The situation changed considerably during the 1990s. Due to the increase in ODA volume and the rapid appreciation of its currency, Japan became the largest donor country in the world. As the top donor, it needed a philosophy to underpin its foreign-aid practices. In addition, the end of the Cold War had significantly changed the international context. Instead of engaging in East–West competition, every country was expected to contribute to international peace and security. This represented an unprecedented challenge for Japan. Since its disastrous defeat in the Second World War, the country had avoided addressing politically sensitive issues, including international peace and security. Instead, Japan had concentrated on economic development, under the dominant influence of the U.S. There was a broad consensus among members of the public that Japan should never fight another war in any part of the world.[6] In addition to its constitutional restriction, which established a high bar for using the Japanese security forces to preserve international peace,[7] Japan had a cardinal principle that strictly prohibited the export of arms or weapons. In 1991, the Gulf Crisis forced Japan to reflect seriously on the contribution it could make to international peace and security.[8]

In short, the new international environment and Japan's position as the world's top donor obliged it to seek a philosophical foundation for diplomacy and foreign aid. One of the most pressing issues was how it could contribute to international peace. Japan's decision to participate for the first time in the UN peacekeeping operation in Cambodia was taken in this context, but, simultaneously, the question of how to deal with the issue of peace-building remained unresolved.

Between the 1990s and the mid-2000s, Japan issued two ODA charters (Government of Japan, 1992, 2003). In contrast to the MOFA (1981), both were officially approved by the cabinet. The first charter was drafted immediately after the Gulf Crisis. Although the Crisis undeniably triggered the charter-drafting process, discussions about ODA reforms had begun in the mid-1980s, during a period of severe criticism of the Japanese government for scandals involving ODA projects.[9] The charter

[6] Due to Japan's devastating experiences during World War II, as well as serious reflections on that war, this sentiment continues to be very strong among the Japanese people.

[7] The constitution of Japan stipulates that, "land, sea, and air forces, as well as other war potential, will never be maintained" (Article 9). This is why the official name of the country's security forces is the "Japan Self-Defense Forces (JSDF)." There are regularly fierce debates over JSDF's participation in international peacekeeping operations. The adoption of the "five-point principle on PKO participation" (see below) is one result of these debates.

[8] Following the Iraqi invasion of Kuwait and the Gulf War, Japan was placed under strong international (particularly, U.S.) pressure to contribute to the Coalition forces. Although Japan made a huge financial contribution of 13 billion dollars, countries like the U.S. and Kuwait appreciated this much less than expected.

[9] The first major ODA scandal, revealed in 1977, involved the Yen loan project to construct the Seoul subway system. After the downfall of President Ferdinand Marcos in 1986, the prosecution conducted an extensive investigation of ODA projects in the Philippines, on suspicion of bribery. During this period, when the volume of ODA was rapidly increasing, the image of aid was somewhat tarnished in Japanese society. Against

officially presented Japan's philosophy, emphasizing humanitarian considerations, interdependence, and the responsibility of a great power for maintaining international peace and stability, alongside the basic principle of assisting the efforts of developing countries to advance their own economic development. It also established a new criterion for the provision of ODA: steps toward democratization and a market-oriented economy.

The most important issue in this first charter involved the non-military use of ODA (Araki, 2020, 189). Given the heated debate over international cooperation to ensure peace and security following the Gulf Crisis, and based on the four ODA guidelines formulated in 1991,[10] the charter laid down a set of principles for providing ODA in order to avoid the need for military intervention and prevent international conflicts. This showed what the government considered to be its first priority. First, it wanted to clarify that ODA had nothing to do with military assistance. At this stage, however, it was still unclear how Japan would use ODA to contribute to international peace.

The second ODA charter, which was drafted after long and extensive discussions, reflected the development of international circumstances since the mid-1990s. The following three points deserve mention. First, the second charter made poverty reduction the top priority; closely linked with the DAC (1996) and the Millennium Development Goals (MDGs), over which Japan had some ideological influence, given the country's significant amount of ODA funds.[11] Second, peace-building was mentioned as one of the four priorities. This was Japan's first charter to mention this idea. Since the mid-1990s, following the United

this backdrop, the opposition parties brought ODA bills before the parliament several times during the 1980s.

[10] In the guidelines, the Japanese government clarified four issues that had to be considered when providing ODA. Three of the four issues dealt with military matters: military expenditure, the development of the weapons of mass destruction, and arms trading. Japan declared that it would limit development assistance to countries engaged in these activities. As positive factors, Japan considered the advance of democratization and a market-oriented economy when allocating ODA.

[11] A Japanese diplomat (the former ambassador to Vietnam) reported that Japan took the overall lead in discussions about drafting DAC (1996), which built the foundation for the MDGs (Hattori 2003). Although his statement may have been an exaggeration, it would make sense for Japan, as the top donor country, to have had significant influence in the DAC at that time.

Nations (1992), several donor countries had begun to engage in peace-building activities, including disarmament, demobilization, and reintegration (DDR), security-sector reform (SSR), and refugee assistance. Third, the idea of human security was introduced to the charter as a basic viewpoint. The charter presented human security as a policy concept closely related to the peace-building agenda.

2.3 The Mid-2000s: Age of realism

There have been tremendous changes in the international situation since the 2000s. Politically, the mid-2000s saw the proliferation of attacks by religious extremists and the retreat of democracy all over the world. There were also contrasting economic trends: while emerging economies, such as China and India, recorded significant growth, developed economies were generally weak and unstable. These trends culminated in the financial crisis of 2008–2009. One consequence was a decline in the importance of DAC ODA that flowed to developing countries, in comparison to funding from other sources, such as private-sector investment and development (e.g., the Bill and Melinda Gates Foundation). New non-DAC donor countries also emerged, including China, India, and the Gulf countries (Severino & Ray, 2009). As a result of these changes, DAC donors began to take a realistic stance on ODA policy.

Japan was no exception. The rise of China had a strong impact on Japan, focusing attention on national security.[12] Japan's long-term economic stagnation since the 1990s and the effects from the Great East Japan Earthquake in 2011 made it difficult to increase the volume of ODA and reinforced the tendency to use the funds for its own economic recovery. In addition, the government had been under pressure from the business community since the 2000s (Watanabe, 2019). As a consequence of the drastic reduction in tied loans since the end of the 1970s, Japanese private companies were finding it increasingly difficult to obtain any benefits from the ODA projects. The business community demanded that ODA policy be revised for the sake of the national interest (Nihonkeizaidantairengoukai, 2003).

In 2015, the Development Cooperation Charter was drafted to account for these demands. While the Charter clearly expressed "the

[12] Amid high tensions with China, the National Security Strategy was drafted in 2013.

basic stance of Japan to earnestly tackle challenges facing the international community," the most important change was the reference to "national interests." For example, in the section titled "philosophy" in the charter (Government of Japan, 2015), it is stated "(s)uch cooperation will also lead to ensuring Japan's national interests." The national interests defined in the Charter were those of Japan and the international community.[13] However, it is undeniable that the Charter obliged those in charge of Japan's ODA to take the national interest into account in their practices. This consideration of the national interest is now explicitly institutionalized in the ODA evaluation process, which systematically investigates whether projects, schemes, or policies have contributed to this objective (MOFA, 2020).[14]

3 Policy Concepts and Their Transformation

Generally, policy concepts are established by the government to provide a philosophical basis for concrete practices. Since the 1990s, Japan has actively used some policy concepts to clearly indicate its normative rationale for foreign aid. The use of these concepts deserves careful examination, as they reflect not only Japan's efforts to present the characteristics of its ODA policies, but also the gap between its stated philosophy and reality. This section analyzes three policy concepts: human security, TICAD, and FOIP.

3.1 Human Security

Human security has been one of Japan's most important policy concepts, not only in foreign-aid policy, but also in diplomacy in general. In fact, Prime Minister Obuchi (1998–2000) took active steps to promote this concept in Japanese diplomacy. The most notable action in the international context was the establishment of the United Nations Trust Fund

[13] As examples of national interests, the charter lists the following: "maintaining its peace and security, achieving further prosperity, realizing an international environment that provides stability, transparency and predictability, and maintaining and protecting an international order based on universal values" (Government of Japan, 2015, 3).

[14] Currently, Japanese ODA is evaluated from a "development perspective" and a "diplomatic perspective." The former is based on the DAC principle, and the latter on the national interest (MOFA 2020).

for Human Security in 1999. The idea was adopted as a basic principle in both the 2003 ODA Charter and the 2015 Development Cooperation Charter and was repeatedly mentioned in the 2013 National Security Strategy. It is safe to say that Japanese diplomacy has adopted this concept as a critical component.[15]

The Japanese government has been actively involved in shaping human security. It is widely recognized that the concept is composed of two main pillars, "freedom from fear" and "freedom from want."[16] Since the concept began to attract worldwide attention in the mid-1990s, several countries have adopted it as a philosophical rationale for their diplomatic policies. Although Canada and Japan were both very eager to promote it, they emphasized different aspects of the concept, with Canada emphasizing "freedom from fear" and Japan "freedom from want." In line with this, the Canadian government made a commitment to shape the idea of "Responsibility to Protect (R2P)" (ICISS, 2001). Considering its own field of expertise[17] and anticipating opposition from developing countries, Japan took a different stance and advanced its agenda within the arena of the United Nations. One result was the UN (2010) resolution, in which human security was defined as a concept resembling human development[18] and respecting the state sovereignty and was clearly different from the R2P. The essence of this report was included in a resolution of the UN General Assembly two years later (UN, 2012). In this way, Japan succeeded in mainstreaming its interpretation of human security in the United Nations General Assembly.

Currently, human security tends to be understood as something similar to human development among Japanese stakeholders (Kurusu, 2016,

[15] See also (Kurusu & Kersten, 2011). For instance, JICA defines its mission as follows: "JICA, in accordance with the Development Cooperation Charter, will work on human security and quality growth" (https://www.jica.go.jp/english/about/mission/index.html, accessed on March 21, 2021).

[16] UNDP 1994. Later, the third pillar, the "freedom to live in dignity" was added (UN, 2010).

[17] The idea of "freedom from want" was (and is) very familiar to Japan's ODA community. In addition to its original focus on economic cooperation, Japan has offered cooperation in the fields of education and public health since the 1960s. Even before the Japanese government officially adopted the idea of human security, its ODA began to emphasize human development.

[18] The report states, "(h)uman security and human development are thus two sides of the same coin" (UN, 2010, para 15).

20). In other words, the understanding has a clear bias toward issues relating to "freedom from want." Strictly excluding the restriction of state sovereignty and the use of coercive force in the name of human security, the understanding seems to omit an important aspect of "freedom from fear" (Kurusu, 2011). This may be expedient for Japan, which has limited experience of peace operations because of its constitutional restrictions and the strong reluctance of the general public to engage with this issue. However, this raises the following question: does human security without "freedom from fear" make any sense in today's world, in which complex armed conflicts proliferate? Although the UN resolution has legitimized this interpretation of human security, removing an important part of the concept seems to have made it significantly less attractive.

3.2 Tokyo International Conference on African Development (TICAD)

Since its first meeting in 1993, the Japanese government has periodically organized a summit with African heads of states under the name of TICAD. The conference was held once every five years until 2013 after which the interval was shortened to three years. The next TICAD VIII will be held in Tunisia in 2022. Although "TICAD" is literally the name of the conference, this paper also considers it to be a policy concept. MOFA planned the first TICAD conference in 1993 to promote its engagement in Africa. It is therefore appropriate to regard TICAD, not simply as the name for a series of events, but as a Japanese policy framework for Africa. In fact, each TICAD clearly reflects the current trend in Japan's Africa policies.

The idea for the first TICAD in 1993 was developed during rapidly changing international circumstances. In addition to the requirements of its position as the top donor country, Japan urgently needed a guiding philosophy in Africa. The important context was its relationship with South Africa.[19] In 1988, Japan had been publicly criticized at the UN General Assembly for expanding trade with South Africa under Apartheid. TICAD emerged and was approved within MOFA as a measure to enable sanctions against South Africa to be lifted; although President de Klerk made serious efforts to democratize the country, the ANC leader, Nelson

[19] This section is based on Shirato (2020).

Mandela, asked the international community to maintain the sanctions. Japan had decided to lift sanctions against South Africa gradually, but it also wanted to make a firm commitment to African development, against a backdrop of aid fatigue among Western donors.[20] The idea for the first TICAD was unique because it was an "intellectual conference" at which donors would not be asked to pledge aid to projects; instead, the objective was to formulate innovative measures to promote African development without evoking images of a rich country throwing gifts to poor nations. Some of the ideas presented at the first TICAD, such as "respecting African ownership and collaborating through partnership" and "conveying Asian experiences to Africa" later became important components of Japan's policy of ODA for Africa.

More than a quarter of a century since TICAD was launched, its features have clearly changed. With hindsight, the turning point seems to have been TICAD IV in 2008, when its business orientation began to dominate. Recently, TICAD has strengthened its role as a business forum, helping private Japanese companies expand their activities to Africa. As organizing business-oriented summits with African countries has become a global trend,[21] the shift in Japan's TICAD policy has been pronounced.

3.3 *Free and Open Indo-Pacific (FOIP)*

Officially announced in 2016, FOIP is a relatively new policy concept in Japanese diplomacy. The concept clearly highlights Japan's diplomatic philosophy and links with ODA projects. FOIP is composed of three pillars (Government of Japan, 2019). The first pillar emphasizes maintaining the fundamental principles of the international order, including the rule of law, freedom of navigation, and free trade. The second pillar, titled "pursuit of economic prosperity," expresses a strategy of economic development by enhancing connectivity. It aims to improve physical connectivity through "quality infrastructure" and also to strengthen people-to-people connectivity, such as education and training, and institutional connectivity, including harmonization and common rules such

[20] The volume of total ODA funds for African countries declined after the end of the Cold War and stagnated during the 1990s.

[21] TICAD-style summits with Africa have been organized by various countries, including China, India, Russia, South Korea, Turkey, and the UK. The main purpose of all these summits has been to promote business.

as EPA/FTA. The third pillar demonstrates a commitment to peace and stability through capacity-building, humanitarian assistance, and disaster relief.

Like TICAD, FOIP is the product of rapidly changing international circumstances since the 2000s, in particular, the growth of China. This policy concept was formulated under Prime Minister Shinzo Abe during a period of increasing tension with Japan's giant neighbor. During the late 1990s, the Obuchi administration attempted to activate the ASEAN + 3 framework to improve its relationship with China and South Korea. Although this diplomatic policy was pursued in the 2000s under several prime ministers, it was replaced in the 2010s with FOIP, based on a new perspective, which focused on the "sea" rather than the "region" (Yamakage, 2016, 226).

Abe became Prime Minister for the second time in late December 2012, when the China-Japan relationship was extremely tense, due to a territorial dispute over Senkaku Island. His firm stance against China was clearly shown in his article, "Asia's Democratic Security Diamond" (Abe, 2012), published just after his inauguration. In the article, he called for the creation of a "Democratic Security Diamond," in which Japan, the U.S., India, and Australia would cooperate to secure maritime commons, stretching from the western Pacific region to the Indian Ocean, based on a keen suspicion that China had strengthened its expansionist policy toward the East and South China Seas. Although official announcements of FOIP were made in Nairobi at the time of TICAD VI, in August 2016, the basic idea was developed much earlier.[22]

While the FOIP announcement was Japan's response to the new international circumstances, it is important to note that the policy concept has been transformed since its inception. As an expert on Japanese diplomacy has pointed out, the major objective of Japan's FOIP strategy remains quite ambiguous (Oba, 2018, 13). One of the reasons for FOIP's ambiguous strategy is the fact that Japan's stance has not been consistent in reflecting its diplomatic relationship with China. As the concept of the "Democratic Security Diamond" reveals, the relationship was very strained at the time of Abe's second inauguration in 2012. However, it has gradually improved since then, to the extent that Abe

[22] Abe's speech to the Indian parliament in August 2007 is often referred to as the origin of FOIP (Suzuki 2017, 133). It was delivered during an official visit, just before he resigned the position of prime minister in September 2007.

even promised to cooperate with China's Belt and Road Initiative in his policy speech at the 196th Diet in January 2018. As the bilateral relationship improved, the Japanese government mentioned FOIP less frequently and loudly. However, the bilateral relationship deteriorated again during the Covid-19 pandemic for various reasons, including increased U.S.-China tension, China's new maritime-security law, and Chinese actions in the Senkaku Islands, South China Sea, and Hong Kong. Prime Minister Yoshihide Suga, who assumed his post in September 2020, has emphasized FOIP on several occasions.[23] Although FOIP is a policy concept that connects ODA projects with Japan's diplomatic philosophy, its usage by the government has wavered as Japan's bilateral relationship with China has fluctuated.

4 Outcomes and Challenges

Addressing criticism of its method of providing ODA, Japan began to search for a philosophical framework for aid. Since the 1990s, the country has made consistent efforts to clarify its normative rationale for providing ODA through a series of charters and policy concepts. In recent years, Japan's ODA philosophy has become much clearer. A MOFA booklet published in 1981 explained its role in relation to external pressure. By contrast, the 2015 Development Cooperation Charter vowed that Japan would contribute actively to the international community through ODA and other means. It is important to acknowledge this change.

Critics may argue that the essence of the Development Cooperation Charter lies not in its official statements about the philosophical foundation for aid, but in the term, "national interests," which has been newly inserted in this charter. They may claim that the true purpose of the charter is to benefit Japanese companies, just as ODA did during the 1950s and 1960s. Although there is an element of truth in this argument, the 2015 charter cannot be regarded as a simple atavism, considering the significant broadening of the areas and actors engaged in Japanese ODA since that time. The charter clearly recognizes that national interests should include contributions to the peace, security, and prosperity of the international community. Nevertheless, there will always be a danger

[23] The concept has been mentioned and emphasized by Mr. Suga on various occasions, including his first general policy speech in October 2020, his New Year's policy speech in January 2021, and a US-Japan joint leaders' statement in April 2021.

that national interests, in a narrow sense, will be prioritized. To avoid this, Japan's contributions to national interests in a wider sense should be regularly assessed during ODA evaluations.

Since the 1990s, Japan has promoted several important policy concepts, linking the philosophy of aid with everyday practices. The government has made huge efforts to formulate and disseminate these concepts. Despite the remarkable progress, the concepts have been strongly influenced by real politics, which has often caused conceptual transformations, as revealed in the previous section.

Such conceptual changes have been most clearly observed in relation to human security. The country played a crucial role in mainstreaming a particular interpretation of human security, through a UN General Assembly resolution (UN, 2012), injecting a strong bias in favor of "freedom from want." Although logically, "human security," which is mentioned in the 2003 ODA Charter, implies both "freedom from want" and "freedom from fear," the latter goes largely unmentioned in the 2015 Development Cooperation Charter. The extent to which such a concept can be effective in dealing with today's complex security problems is questionable.

Intentionally or not, this conceptual change reflects Japan's changing position on peace policy. From the late 1990s to the beginning of the 2000s, when DAC donors increased their engagement in promoting the development and stability of conflict-affected countries, Japan was ready to make a contribution. After the Gulf Crisis, the country began to assist the UN peacekeeping operation by sending its Self-Defense Forces (JSDF) to post-conflict countries including Cambodia, Mozambique, and Timor-Leste. In 2001, the first Japan International Cooperation Agency (JICA) guidelines for peace-building were published, following an eighteen-month research project (JICA, 2001). As a consequence of these practices and arguments, the 2003 ODA Charter included peace-building as one of its four priorities.

Although peace-building was similarly mentioned in the 2015 Development Cooperation Charter, Japan has recently become less enthusiastic about this issue than it was during the time of the ODA Charter in 2003. The reasons for this can be attributed to both universal and specific factors. First, the characteristics of the "New War" (Kaldor, 1999) have made peace-building practices more complex and difficult. These difficulties have compounded, due to the expansion of religious extremism. In addition, increasing security concerns have placed severe constraints on Japan's peace-building activities. Concerned about the safety of Japanese nationals, the country has imposed strict restrictions on working conditions in conflict-affected areas. Consequently, most peace-building activities have been conducted exclusively by a handful of government officials and experts, and NGO participation has been significantly limited, severely impeding the development of expertise in this field.

In a complex way, these safety restrictions echo those of the JSDF. Due to the "five-point principle on PKO participation," the JSDF cannot participate in peacekeeping operations without the "existence of a cease-fire agreement between parties to a conflict."[24] This was the main reason for the withdrawal of the JSDF from the UN peacekeeping operation in South Sudan (UNMISS) in 2017, which was seen as a major setback for Japan's peace policy. South Sudan was "too dangerous" for the JSDF. Although the five-point principle was initially developed to prevent the use of arms overseas, it has ultimately made it extremely difficult for the JSDF to contribute to peacekeeping activities, the nature of which has changed significantly, reflecting the complexities of recent armed conflicts (UN, 2000). In this context, the definition of human security adopted by the UN resolution in 2012 seems to have corresponded with and legitimized Japan's reluctance to engage in peacekeeping and peace-building operations.

TICAD, another important policy concept, has also been transformed since its emergence. Although it began as an intellectual conference discussing strategies for African development, it currently resembles a business forum, promoting trade and investment. Every TICAD final statement mentions cooperation, not just to strengthen socio-economic development, but also to promote good governance and peace-building as priority areas. However, it is undeniable that TICAD's central focus is clearly the former. Currently, TICAD faces two major challenges. First, despite the active promotion, Japanese trade and investment in African countries have not increased as expected. For reasons that include a lack of information and experience, different rules, and security problems, Japanese companies remain reluctant to work in Africa. Second, the peace-building and governance components included in TICAD's action plans

[24] The principle was incorporated into the 1992 law on the PKO (Act on Cooperation with United Nations Peacekeeping Operations and Other Operations); no substantial revisions have been made since then.

have produced only limited results. While TICAD began as an innovative device for promoting Japanese diplomacy in Africa, it is no longer clear that this type of conference can produce substantial outcomes.

In the case of FOIP, Japan's usage of the concept has fluctuated, reflecting the country's changing diplomatic relationship with China. Although the policy concept clearly expresses principles such as the rule of law, freedom of navigation, and free trade, the interpretation of these terms has changed in line with specific political circumstances. In general, the Japanese government tends to emphasize FOIP when the bilateral relationship with China is strained. Like other policies formulated under the Abe administration (Envall, 2020), the nature of Japan's FOIP can be characterized as both realistic and pragmatic. The transformation of a policy concept is not in itself problematic. However, if the concept comes to lack consistency, it cannot be easily used to link the philosophical framework and practices of ODA.

5 Conclusion

Over the past several decades, Japan has made constant efforts to establish an ODA philosophy. These efforts were initially propelled by severe pressure from DAC peers, which accused Japan's ODA policy of commercialism. In addition to criticism from abroad, rapidly changing international circumstances, following the end of the Cold War, made it necessary for Japan to clarify and express the philosophical basis for its ODA and diplomacy in general. The two ODA Charters in 1992 and 2003, as well as the Development Cooperation Charter in 2015, were the results of these efforts. After the 1990s, Japan advocated several policy concepts, such as human security, TICAD, and FOIP, that linked its philosophical stance with everyday ODA-related practices. The analysis presented in this chapter reveals that these policy concepts have been transformed over time. The conceptual changes discussed above suggest that Japan has used these concepts rather flexibly, for its own benefit. In other words, the way these concepts have been used reveals the pragmatic nature of Japanese diplomacy.

Pragmatism is not bad in itself; in fact, it may have been a necessary choice, as perceived by the nation's diplomats, to protect national interests. Since Japan has always been exposed to strong pressure from external actors, including the other DAC countries, the U.S., and China, pragmatic attitudes have been indispensable for protecting its national interests. However, pragmatism cannot answer the question of why Japan should provide ODA. Although foreign aid naturally constitutes one tool in the diplomatic toolkit, Japan's reason for providing ODA cannot be explained simply by analyzing changeable international circumstances. The question of altruism is closely related to another fundamental question of what the Japanese values are.[25] The three charters have not yet provided a clear answer to the question. Despite having established three charters, the Japanese will continue to seek inherent reasons for providing foreign aid.

References

Abe, S. (2012). *Asia's democratic security diamond*. Project Syndicate: World Opinion Page. https://www.project-syndicate.org/onpoint/a-strategic-alliance-for-japan-and-india-by-shinzo-abe?barrier=accesspaylog. Accessed 22 March 2021.

Araki, M. (2020). *Kokusai kyoryoku no sengoshi* [Post-war history of international cooperation]. Toyokeizaishimposha.

Arase, D. (1995). *Buying power: The political economy of Japan's foreign aid*. Lynne Rienner.

Development Assistance Committee (DAC). (1996). *Shaping the 21st century: The contribution of development co-operation*. Paris.

Envall, E. D. P. (2020). The 'Abe Doctrine': Japan's new regional realism *International Relations of the Asia-Pacific, 20*(1), 31–59.

Fukuda-Parr, S., & Shiga, H. (2016). Normative framing of development cooperation: Japanese bilateral aid between the DAC and Southern donors *JICA-RI Working Paper*, 130, JICA Research Institute.

Government of Japan. (1992). *Seifu kaihatsu enjo taiko* [ODA Charter]. Tokyo.

Government of Japan. (2003). *Seifu kaihatsu enjo taiko* [ODA Charter (revised version)]. Tokyo.

Government of Japan. (2015). *Kaihatsu kyoryoku taiko. Tokyo* [Cabinet decision on the Development Cooperation Charter (provisional translation)]. https://www.mofa.go.jp/mofaj/gaiko/oda/files/000067701.pdf. Accessed 11 March 2021.

Government of Japan. (2019). *Towards free and open Indo-Pacific*. https://www.mofa.go.jp/files/000407643.pdf. Accessed 22 March 2021.

[25] Fukuda-Parr and Shiga (2016) is one such attempt for searching normative rationale of Japan's ODA.

Hattori, N. (2003). *OECD・DAC 'kaihatsu senryaku' no kaisetsu* [Comments on the OECD-DAC's 'Development strategy']. http://www.devforum.jp/archives/articles/. Accessed 7 March 2021.

Hedberg, H. (1970). *Nihon no chosen: 1980 nendai no keizai chotaikoku* [Japan's challenge: Economic superpower in the 1980s]. Mainichi Shinbunsha.

Hume, D. (2016). *Should rich nations help the poor?* Polity Press.

Independent Commission on International Development. Issues 1980. *North-South: A Programme for Survival.*

International commission on intervention and state sovereignty (ICISS). (2001). *Responsibility to protect.* Ottawa.

Japan International Cooperation Agency (JICA). (2001). *Jigyo senryaku chousa kenkyu: Heiwa kouchiku* [Strategic research for operation: Peace-building]. JICA.

Kaldor, M. (1999). *New and old wars: Organized violence in a global era.* Stanford University Press.

Kurusu, K. (2011). Gendankaino 'ningenno anzenhoshou' ['Human security' at the present stage]. *Kokusaimondai [International Issues], 603*, 5–14.

Kurusu, K., & Kersten, R. (2011). Japan as an active agent for global norms: The political dynamism behind the acceptance and promotion of 'human security.' *Asia-Pacific Review, 18*(2), 115–137.

Kurusu, K. (2016). Does the concept of human security generate additional value? An analysis of Japanese stakeholder perceptions (JICA-RI Working Paper No. 122). JICA Research Institute.

Ministry of Foreign Affairs. (1981). *Keizai kyoryoku no rinen: Seifu kaihatsu enjo ha naze okonaunoka* [Philosophy on economic cooperation: Why do we carry out economic cooperation]. Tokyo.

Ministry of Foreign Affairs. (2020). *ODA evaluation guideline* (in Japanese). https://www.mofa.go.jp/mofaj/gaiko/oda/files/000486015.pdf. Accessed 12 March 2021.

Nihonkeizaidantairengokai (2003). *ODA taiko minaoshini kansuru iken* [Opinions as to the revision of the ODA Charter] https://www.keidanren.or.jp/japanese/policy/2003/033.html. Accessed 26 March 2021.

Oba, M. (2018). Nihon no 'Indo-Taiheiyo' koso [Japan's 'Indo-Pacific' vision]. *Kokusai-Anzenhosho [The Journal of International Security], 46*(3), 12–32.

Pearson, L. B. (1969). *Partners in development: Report of the Commission on International Development.* Praeger.

Severino, J. M., & Ray, O. (2009). The end of ODA: Death and rebirth of a global public policy (Working Paper No. 167). Center for Global Development.

Shimomura, Y. (2020). *Nihongata kaihatsu kyoryoku no keisei: Seisakushi I, 1980 nendai made* [Policy history I: Evolution of the Japanese model of development cooperation, 1945–89]. Tokyo Daigaku Shuppankai.

Shirato, K. (2020). *Afrurika kaihatsu kaigi (TICAD) tanjo hiroku (I)~(VII)* [The secret records on the birth of TICAD], *Shinchosha Foresight*. https://www.fsight.jp/category/ticad. Accessed 27 February 2021.

Suehiro, A. (2005). *Keizai saishinshutsu heno michi: Nihonno tai tounanajia seisaku to kaihatsu taisei* [The way for economic readvance: Japan's policy for Southeast Asia and the development regime]. In. M. Nakamura, A. Amakawa, K. Yun, and T. Igarashi (Eds.), *Sengo kaikaku to sono isan [Post-war reform and its legacies]*. Iwanami Shoten.

Suzuki, Y. (2017). *Nihon no senryaku gaiko* [Japan's strategic diplomacy]. Chikumashobo.

United Nations. (1992). *An agenda for peace: Preventive diplomacy, peacemaking and peace-keeping* (A/47/277–S/24111).

United Nations. (2000). *Report of the Panel on United Nations Peace Operations* (A/55/305–S/2000/809).

United Nations. (2010). *Human security, Report of the Secretary-General* (A/64/701).

United Nations. (2012). *Resolution adopted by the General Assembly on 10 September 2012. 66/290.* Follow-up to paragraph 143 on human security of the 2005 World Summit Outcome. (A/RES/66/290).

United Nations Development Programme. (1994). *Human development report*. Oxford University Press.

Watanabe, M. (2019). *Henkasuru kankyoto nihonno enjoseisaku: Futatsuno seifukaihatsuenjotaikono sakuteikara* [Changing environment and Japan's aid policy: Establishment of the two development cooperation charters] (Background Paper No.4). JICA Research Institute.

White, J. (1964). *Japanese Aid*. Overseas Development Institute.

Yamakage, S. (2016). *Nihon no chiiki koso to 'Chugoku no taito' – Rekidaishusho no seisakuenzetsu ni miru 'nakama' no egakikata* [Japan's vision for regionalism and the 'rise of China': How Japanese prime ministers have described their 'friends']. In. M. Oba (Ed.), *Higashi Asia no katachi: Chitsujo keiseito tougouwomeguru Nichibeichu ASEAN no kosa [The making of East Asia: Order, integration and stakeholders]* (pp. 205–237). Chikura shobo.

Yasutomo, D. T. (1986). *The manner of giving: Strategic aid and Japanese foreign policy*. Lexington Books.

CHAPTER 2

Reflection on a Normative Rationale for Korean ODA Policy: Duty, Self-Regards, and Obligation

Huck-ju Kwon

1 Introduction

Despite economic growth and global fights against poverty, a large number of people in the world still suffer not only from poverty but

An earlier version of this contribution was presented at the International Conference of the Asia Social Science Research Council in Hanoi, Vietnam, on 24–25 September 2019. This research was supported by the 2019 Basic Research Grant from the Asia Development Institute, Seoul National University.

H. Kwon (✉)
Asia Development Institute, Seoul National University, Seoul, South Korea
e-mail: hkwon4@snu.ac.kr

Ash Center for Democratic Governance and Innovation, Harvard Kennedy School, Harvard University, Cambridge, MA, United States

H.-j. Kwon et al. (eds.), *International Development Cooperation of Japan and South Korea*,
https://doi.org/10.1007/978-981-16-4601-0_2

also hunger and disease. Increasing numbers of natural disasters aggravate problems caused by social conflicts in many developing countries. The global coronavirus pandemic, which will further exacerbate global poverty, is an unfortunate but vivid example. It is true that global efforts such as the UN's Millennium Development Goals and Sustainable Development Goals have made great impacts over the last few decades. The number of extremely poor was halved from 1.9 billion in 1990 to 836 million in 2015 (United Nations, 2015). Nevertheless, tremendous challenges for development remain at the global level. Economic recession during the global pandemic will inevitably increase the number of poor people across the world. Preliminary estimates predict that poverty at the global level could increase by half a billion people, or 8% of the global population (OECD, 2020). The majority of people in extreme poverty will be in developing countries. In many fragile places in the world, such as Syria, Afghanistan, Libya, and Myanmar, where innocent people are killed, injured, and displaced, people's livelihoods will further deteriorate. The global pandemic shows that the global community should work together to fight against common enemies, such as viral epidemics and climate change, but it has turned out to be difficult to mobilize global efforts as some countries put their national interests first instead of the global common good. It will be a great challenge to find common ground for such efforts. Global efforts should be mobilized not only with the motivation of economic advantages but also on ethical grounds. In this context, this chapter will reflect on the normative rationale for ODA policy in the Republic of Korea (hereafter Korea) and search for ideas for the future direction of ODA policy. For such purposes, it will explore different strands of normative rationales for international development and critically analyze the underlying rationale of Korean ODA.

2 Reflexive Approach to ODA Through Self-Awareness

With the increasing role of East Asian countries, including Korea, in global efforts for poverty reduction and development, an important question has emerged about their policy rationale for international development cooperation. Are East Asian countries using international development for their own national interests, particularly economic and commercial advantage, or are they really pursuing desired goals such as supporting human freedom, reducing poverty, and creating the global public good? A body of literature which looks into the characteristics of international development cooperation in these countries tends to show that commercial advantage has been the main priority in international development cooperation in Japan and Korea (Lancaster, 2007; Stallings, 2017). In 2018, according to the Commitment to Development Index, Korea was at the bottom among 27 countries, while Japan was ranked in 24th place (Center for Global Development, 2019). While Norway, Denmark, and the Netherlands, as well as Sweden, are seen as high-quality donors, Korea and Japan are seen as doing very little in terms of the contribution to poverty reduction and social development in developing countries (Hulme, 2015). What are Korea and Japan doing in terms of their international development cooperation if their commitment to development has been so low?

Such questions can be answered through empirical investigation, and some studies argue that Korean ODA has been strongly influenced by economic considerations (E. M. Kim & Oh, 2012). In this book, Kim and Song's chapter clearly shows that Korean ODA policy has been influenced by foreign policy goals, including Korea's economic interests. Nevertheless, questions about the purpose of Korean ODA do not have a foregone conclusion. If the questions about our reasons for international development cooperation are put to ourselves as Korean citizens, different answers may be given from those based on economic considerations. Why should we, Korean citizens and the government, support developing countries to pursue our own interests before their development? How should we implement international development policy if we commit ourselves to international development cooperation? These questions should be answered based on the normative values and ideas we have, and moral and ethical reflections are required for us to answer. This chapter aims to shed light on the normative aspects of Korean ODA policy.

Reflection on such questions is particularly pertinent at this point of time. Although Korea is still seen as an emerging donor, it is now time to think about providing ODA as a mature donor since it has been 10 years since Korea became an official member of the OECD/DAC. For the first decade since its membership, Korea has tried its best to follow international rules and guidelines, including the OECD/DAC guidelines (Stallings, 2017). The administrative structure of ODA has been reformed and the size of ODA has steadily increased, following the recommendations of the OECD peer reviews (OECD, 2018). This was understandable for a country which began implementing ODA as a new member. However, there has been increasing criticism that Korea lacks a clear sense of purpose and vision in international development policy.[1] This lack of a common vision and sense of purpose can easily lead to fragmentation in the policy system of international development cooperation. If the various policy actors involved do not share a clear vision or goal for international development cooperation, they will pursue paths in different directions. Entering the second decade as an OECD/DAC member, Korea needs to come up with a clear sense of purpose for its ODA policy, which should be consistent with the global norms of foreign development assistance. It should also be conceived and decided based on conscious reasoning, not by historic events or by myopic interest calculations.

In the following sections, this chapter will review Korean ODA policy through a reflexive approach. The term reflection implies not only to think about something, but also to be aware of oneself as a subjective agency in relation to the matter concerned (Bourdieu, 2004). In other words, it can be "construed as a form of self-awareness of own perspective" on a subjective matter (Hamati-Ataya, 2013). Reflection also requires contemplating values in addition to the fact since our recognition of fact is inherently linked to the normative dimension of the implicit or explicit paradigm we may have. It is also crucial to reveal the implicit ethical dimensions of our own thinking. Reflecting on the rationale of ODA may lead to the self-discovery of Korea's ODA as well as conscious efforts to conceive ideas and make decisions for future ODA policy.

With such reasoning, this chapter will first discuss different theoretical and normative justifications for international development cooperation, especially in the context of ODA. According to theories from the realism school of international relations, states on the international stage pursue national interests, and international politics can be like anarchy, with each state struggling against the others for survival (Waltz, 1979). ODA as part of foreign policy will be nothing more than a policy instrument for the national interest. However, from the perspective of international society, states are members of the international community and should follow certain rules and norms of action (Bellamy, 2011). ODA could be a policy instrument to follow rules and norms for the benefit of the international

[1] Interview with Professor Eunmee Kim, a leading expert in the field of Korean ODA. The interview was conducted on 5 November 2019.

community. In reality, states follow paths between purely national interests and global rules and norms. This is because of international politics, which is based partly on power politics and partly on international order. It is also necessary to consider the domestic politics of donor countries. In particular, in liberal democratic societies including Korea, there is a large section of society demanding that the government follow the rules and norms of the international community as well as people demanding the pursuit of national interests.

In light of ODA, this chapter will first explore the notion of moral duty, which refers to a duty drawing on individual internal moral values, in contrast to the notion of ethical obligation, which will refer to compulsory requirements stemming from the relationship between members of the community (Harsanyi, 1982). This chapter will also examine the notion of self-regarding rationale in order to refer to the considerations of self-respect as well as diplomatic and economic self-interest (Lancaster, 2007).

Secondly, it will attempt to reflect on the normative grounds of Korean ODA policy and discuss its background and characteristics. Korean ODA began in the early 1960s in the form of training programs for officials from other developing countries, but it really started in 1987 when the Korean government established the Economic Development Cooperation Fund (EDCF). In 1991, the Korean International Cooperation Agency (KOICA) was established, and in 2010, Korea became an OECD/DAC member as an official donor country. Each historic event had its own background and motivation. This chapter will look into such historical events and junctures in order to analyze the motivation and ethical rationale for Korea's ODA in relation to those events.

Lastly, drawing such an analysis and discussion, this chapter will explore the policy rationale of Korean ODA for the future, while incorporating the strength of Korean development experience and knowledge. It will suggest a three-pronged approach to aid for human freedom, aid for socioeconomic and political development, and aid for the global public good.

3 Why Should a Nation Support Other Countries Through ODA?

Morgenthau once remarked that foreign aid is one of the real innovations in modern foreign policy. Foreign aid is not only an instrument of foreign policy to pursue national interests but also an end itself (Morgenthau, 1962). In order to justify international development cooperation as an end itself without resorting to national self-interest, it is necessary to establish its ethical grounds. In his seminal paper, Singer argues that "if it is in our power to prevent something bad happening, without thereby sacrificing anything of comparable moral importance, we ought, morally, to do it" (Singer, 1972: 231). According to this point of view, proximity or distance does not matter if a person needs help, whether this is a neighbor's child or someone in Bengal. It is a moral duty that should be applied to everyone equally without considering nationality (Hulme, 2016), no matter whether the person lives in a Western country or an East Asian country such as Korea or Japan. This simple but powerful argument sets out a moral basis for international cooperation in terms of moral duty. Singer shows that humanitarian work is not charity but duty, and people with such a moral belief should do their moral duty (Singer, 1972).

If we recall the most devastating natural disasters in recent years, for instance the Aceh tsunami in Indonesia in 2004 and typhoon Haiyan in the Philippines in 2013, people from all over the world, no matter whether they lived in neighboring countries or faraway nations, came together to provide relief to those affected. They sought to fulfill their moral duty. Individuals as moral agents should participate in supporting people suffering from poverty and natural disasters in other countries. Nevertheless, people who do not fulfill their moral duty are not forced to do so, but we may praise those who do.

All activities to help other people, however, may not come out of strong moral duty. A weaker version of moral duty is self-regarding consideration. It is a contrasting concept to other-regarding viewpoints, through which a person sees the well-being of other persons as a priority over his or her own well-being (Sen, 1993). By contrast, one may help others out of self-regarding concerns. For instance, we may help others to gain respect from friends or neighbors. While you are walking along the street with your friend, you find a person asking for donations for charity. Even though you may not know exactly what the charity is about, you would still donate some money because it makes you a good person. Such self-regarding considerations may not be bad and are partly based on a moral belief: the premise that helping others is morally desired. It has, however, a serious weakness as a moral act since it does not seriously concern those who are in need, nor does it render the outcome of helping others a main priority; rather, a self-regarding consideration is the primary objective. For this reason, self-regarding acts can only be understood as weak moral behavior.

Now, let's turn to the notion of ethical obligation in contrast to moral duty. Suppose that moral duty comes from an agent's inner morality, and ethical obligations are based on norms from the community that people live in together. Moral duty is weaker than ethical obligation in terms of the extent to which one has to provide support for others in need. Passers-by could certainly help a child who fell into a river, but they may not have an obligation to provide help to those in faraway places suffering from a serious natural disaster. They are not obliged to do so although it is still a moral duty. In terms of justice as fairness, however, it is an ethical obligation to provide support to those in the worst position in the community. According to Rawls, those in the worst position have a legitimate claim to the primary goods necessary for them to live as free and equal individuals (Rawls, 1982). The rest have to provide the necessary resources since they could at some point be in such situations as well. Here it is very important to recognize why the moral duty to support those in need becomes a political obligation. In the Rawls theory of justice, individuals enter into a social contract, which could represent the basis of an institutional framework for a political community, and under such a social contract the moral duty to support those in need becomes a stronger ethical obligation.

Such reasoning, however, is faced with a significant difficulty regarding international development assistance. Extending this line of thinking, Nagel argues that such an obligation can only hold within a nation in which citizens set up a constitutional framework of rights and duties to each other under which they agree to be obliged (Nagel, 2005). A corollary of this argument is that it is not an obligation to help the suffering poor living in distant countries without sharing citizenship. An ethical theory based on individuals sharing a political community as a sole moral agency cannot provide a sufficient theoretical basis for international development assistance although it still has a strong appeal to us to help people in difficult situations.

Further, international development assistance is not only to support those in need, such as people suffering from humanitarian crises and severe poverty, but it is also provided for economic and social development. Even if there is no visible catastrophic humanitarian crisis, billions of people could be suffering from chronic poverty or other livelihood difficulties that prevent them from acquiring capabilities. As Sen defines development as freedom (Sen, 1999), development is a process in which individuals acquire adequate functioning and capabilities to exercise their freedom. In order to exercise freedom, however, people also need to have options to choose and arenas in which they can exercise freedom, which are not necessarily attached to individuals (Ringen, 2007). Political and economic institutions should be in place and civil rights must be protected so that individuals can have options to choose in terms of their own freedom. The social and economic as well as physical environment should establish arenas wherein individuals can implement their life plans. A body of literature in development studies shows that people in many developing countries suffer from institutional and structural barriers to exercising their freedom (Kabeer & Cook, 2000; Midgley, 2014). In a nutshell, freedom hinges not only on individual capabilities but also on the social and economic development of society.

Is it, then, our ethical obligation to provide help to others in distant countries to build social institutions and economic infrastructures that could ultimately enhance people's freedom? It is not intuitively clear that there is a normative link between the ethical obligation of individual citizens and social and economic development in developing countries of which they are not citizens. Here we need to turn our attention to the state as an ethical agency.

As globalization deepens, we as individuals can live our lives in the global arena to a greater extent than in other periods in history. Nevertheless, the global order essentially remains between sovereign states, while other actors such as individuals, social groups, and private enterprises are engaged in global activities by and large with their attached nationality. There is no global government that can define the rights and responsibilities of global citizens, and ordinary citizens should therefore be protected through their sovereign states in the global arena. As global rules and regulations are set up between sovereign states, there are rights and obligations to each other.

The global order has been forced by the powerful developed countries according to Pogge (2008), and it can therefore be said that the plight of people in developing countries under the current global order has been caused by the developed countries. Through this argument, Pogge provides normative grounds for an obligation to support development in developing countries. There are two critical points from this background for our discussion. First, if harm is caused by the global order, which is unfairly imposed on developing countries, it is necessary to reform the global order. International development assistance would only perpetuate an unfair global order without serious efforts to reform it. Secondly, the implication is that a middle-sized country like Korea, which has never had a strong influence on the global order, does not have such an ethical obligation to provide international development assistance. Although it is a salient topic to look into whether the global order is unfair to developing countries, Pogge's theory leaves us a great deal of room for debates about ethical obligations among nations.

Nevertheless, we may not need such a strong argument of unjust global order to establish the obligation of affluent countries for the development of developing countries. Looking back on the history of Korean economic development, for instance, it is very clear that Korea has benefited from the global market and liberal international order for the last 60 years since the 1960s. Korea's economic takeoff is due to a great extent to an export-oriented economic development strategy that was sustained through global free-trade regimes such as GATT and the WTO. In the same period, many sub-Saharan African countries, which were at a similar level of economic development in the 1960s as that of Korea, have not been able to develop their economies. In other words, the global order leaves some countries better off and others worse off. Given the global order within which sovereign states in the world interact with each other, those benefitting from it must have an ethical obligation to support those lagging behind to gain equal footing.

Once the state as an ethical agent enters the field of international development assistance, we can begin to talk about official development assistance.[2] Donor government involvement is necessary to catalyze the social and economic development of recipient countries. As an ethical agent, the state can provide support for the development of public institutions, which represent a critical requirement for sustainable long-term development, on top of humanitarian aid and support for individual freedom. Such efforts include building the institutions necessary for economic activities, the protection of human rights, and political participation. The lesson we can elicit from the developmental state in Korea is the importance of effective public institutions for economic and social development (Adelman, 1997). Since there are drawbacks and historical legacies of the developmental state with authoritarian characteristics, it is not desirable to adopt a wholesale form of the developmental state.

[2] Official development assistance (ODA) is defined as assistance to promote the economic and social development of developing countries financed by donor countries' public expenditure, including grants and concessionary loans.

Democracy as well as effective government are important conditions to ensure individual freedom.

On the domestic front, the state has political responsibility and accountability to its citizens. Once the government supports another country for economic and social development, public expenditure, financed by taxation, is used, and some citizens may argue that public money should be given to a country's own citizens. It is therefore necessary for governments to justify to their citizens that international development assistance such as ODA can be compatible with meeting the requirements of a country's own citizens.

From a perspective of the whole global community, ODA is about not only benefitting other countries but also creating global public good. This is the third factor that lays the normative grounds for official development assistance. ODA can serve donor countries' national interests through the global public good, which would then serve ordinary citizens. It is true that certain forms of tied aid, such as conditions to serve the national interests of donor countries, are not regarded as the desired practice of ODA. It is also a very narrow form of national interest. National interests could, however, be served by ODA, but in indirect ways through creating global public good. For instance, supporting international efforts to combat deforestation through ODA to developing countries in tropical regions will certainly contribute foremost to the global public good, but it also serves the national interests of donor countries since protecting the rainforest would benefit people in donor countries. Fighting climate change by protecting rainforests and oceans is about creating global public good, which will then serve all people equally in the world. As members of the global community, sovereign states, especially the affluent ones, have an obligation to contribute to the creation of the global public good. Serving national interests is not necessarily contradictory to creating the global public good.

4 A Reflection on the Implementation of Korean ODA

This chapter, thus far, has examined various normative grounds for international development assistance including moral duty, self-regarding considerations, and ethical obligation. It has also discussed the ethical obligation of the state as a member of the global community to help other countries and to contribute to the global public good. Given these different strands of normative rationales for ODA, in which direction has Korean ODA gone? In this section, in order to explore this question, the changing rationale of Korean ODA over recent years will be reviewed.

As an emerging donor country, Korea has been perceived as setting a benchmark for developing countries in the international community. For many developing countries Korea provides a unique reference for development policy because Korea was previously one of the poorest countries in the world, and its development experiences remain vivid and relevant to developing countries (Kohli, 2004; Kwon & Koo, 2014). Public support for international development cooperation in Korea has been very solid, as policymakers and ordinary citizens alike are very proud that Korea has become a donor country helping other developing countries. According to the National Survey on ODA in 2017, 86.7% of Koreans supported the provision of ODA. Since the National Survey on ODA was first conducted in 2011, 80 to 89% of citizens have supported ODA for the last decade (Park & Kim, 2017).

As such, Korea has attracted the attention of the international community as an emerging donor within a relatively short period of 10 years. At the global level, Korea has begun to play a significant role in the international development community. In 2011, Korea hosted the Fourth High-Level Forum on Aid Effectiveness in Busan, which called for development effectiveness in terms of the policy orientation of ODA. The Busan High-Level Forum is one of the biggest international policy forums convened in Korea, which symbolizes the significance of Korea's new global status.

A question arising from this observation involves the policy rationale of Korean ODA. Since joining the OECD/DAC in 2010, Korea has implemented two five-year mid-term policy strategies following the Basic Law of International Development and Cooperation. The basic law, revised in 2014, stipulates the primary goals of ODA policy in terms of human prosperity and world peace.[3] It then sets out the following policy goals in relation to developing countries: (1) poverty reduction and well-being improvement, (2) institutional improvement for development, and (3) friendly cooperation and collaboration. The basic law also includes a contribution to solving global problems as one of the main policy goals. Even though these values and policy goals are universally desirable, they

[3] www.law.go.kr; Basic Law of International Development and Cooperation, accessed on 29 October 2019.

still remain at the abstract level. It is necessary to carefully examine the consistency between policy values and goals on the one side and policy practices on the other side in order to understand the meaning of these abstract values and goals.

The mid-term strategic plan involves government documents that are useful for analytical purposes. The first mid-term strategy contains instructions on how to organize the public administrative system for international development assistance in Korea, such as developing detailed contents for a policy program and reorganizing the aid system effectively (Government of Korea, 2010). The second mid-term strategy covering the period of 2016–2020 states the principle of "integrated, substantial, and participative ODA", which is also directed toward the domestic administrative system in Korea (Government of Korea, 2015). The mid-term strategy for international development cooperation policy needs to address approaches of international development policy in order to pursue goals of poverty reduction and the production of the global public good. However, Korea has only established an ODA strategy with policy objectives and targets that are mainly incorporated for domestic consideration, so it could still be considered domestic policy. For this reason, at the stage of implementing international development cooperation, various public institutions within the Korean state do not share common norms and a common sense of purpose; rather, there are different motivations and values pursued by different institutions and individuals. In the following, we will look in more detail at the motivations for Korean ODA revealed in practice.

At the onset, it is necessary to understand why Korea wanted to be a member of the OECD/DAC to understand the underlying idea of ODA policy in Korea. Sixty years ago, Korea was one of the poorest countries in the world; it was a poverty-stricken and war-torn society. After rapid industrialization and subsequent economic growth since the 1960s and 1970s, Korean society became affluent (Song, 1990). In the 1980s, Korea went through a democratization process and was able to consolidate democracy in the 1990s. In terms of social equality, Korea is one of the few countries that has been able to maintain social equality during rapid economic growth (Kwon, 1998). It was a remarkable achievement in terms of economic development with social equity and political freedom. With such an achievement in the background, in 1993 the newly elected civilian government set an open economy in the globalizing world as one of its main policy priorities. In 1996, Korea joined the OECD, an international organization of industrialized countries. It could have been an important landmark for Korea to become one of the leading countries in the world, but it turned out to be a false dawn as the Korean economy descended into turmoil during the Asian economic crisis of 1997/98 and was on the brink of collapse (Weiss, 1999).

Once the economic crisis was overcome, the Korean government once again attempted to regain international recognition of its achievements. Joining the OECD/DAC, which was not completed when Korea became a member of the OECD in 1996, emerged as an important priority for foreign policy in this context.[4] In November 2005, the Korean government formulated a comprehensive ODA improvement plan, which represented the first government policy document regarding ODA strategy (CIDC, 2017). The plan outlined major issues and challenges related to ODA policy in Korea and proposed policy solutions to tackle them, including the creation of a coordinating committee with the government. In particular, it was considered a very important task to coordinate ODA policies of the Ministry of Foreign Affairs and the Ministry of Strategy and Finance, which were responsible for grants and concessional loans, respectively. According to the plan, the Committee for International Development Cooperation (CIDC) was established in 2006 as an inter-ministerial committee chaired by the Prime Minister (CIDC, 2017). In 2008 the Korean ODA system was peer-reviewed by the OECD/DAC, which put Korea's accession to the OECD/DAC on track. After the peer review, the OECD/DAC recommended five points to improve the ODA system in Korea (OECD, 2008):

1. Establishing a legal framework and mid-term strategy for ODA policy;
2. increasing ODA volumes and coordinating aid channels and allocations:
3. strengthening ODA capacity in expertise and evaluation;
4. increasing the effectiveness of aid; and.
5. strengthening humanitarian efforts and cooperation with multilateral institutions.

[4] Interview with former Ambassador Oh Joon, who was responsible for OECD affairs in the 2000s. The interview was conducted on 20 September 2019.

In 2010, Korea was admitted to the OECD/DAC, and it was a very proud and historic moment for the Korean people as it symbolized the global recognition of Korea's development achievement for the last 50 years. A document published in 2010, entitled the Policy for Advancing International Development, which aimed to provide the basic rationale of ODA policy, reflected the government's confidence that OECD/DAC membership would instill great national pride in the Korean public (Government of Korea, 2010). Although poverty reduction and well-being improvement in developing countries were also included as foundational values of ODA policy, the government was determined to maintain this feeling of national pride as the main goal of ODA policy (CIDC, 2017: 11). It was clearly expressed in the speech by then President Lee Myung-bak at the Singapore Summit in 2010. President Lee's feeling was shared by most Koreans, and it showed that ODA policy was strongly underlined by a self-regarding rationale.

> ...a young boy who once stood in the line to receive used clothes from foreign missionaries now stands before you as the President of the Republic of Korea...[Our] remarkable achievement goes out to all those who fought for us and for those who helped us when we were in desperate need...The Korean people who have been through wars and destitution are now prepared to contribute to global peace and prosperity...We are ready to do our part. (from S. Kim, 2011: 812)

Another underlying idea of Korean ODA was to share Korean development experiences with developing countries (Stallings, 2017). It is true that the Korean experience of transitioning from a very poor to an affluent society remains vivid and relevant for many developing countries in designing their public policies. Such an experience is captured by the notion of the developmental state (Woo-Cumings, 1999). The developmental state refers to a policy regime in which economic development is considered the most important policy goal, and a strong state spearheaded by a centralized bureaucracy plays a key role in the policy process. In Ethiopia and Rwanda, for instance, the governments benchmarked the policy regime of the developmental state for their framework of public policy. In Tanzania, the community development movements adopted under the developmental state in Korea in the 1970s, the so-called Saemaul Undong, were experimentally adapted for rural community development projects. To transfer Korean policy experiences in a systematic way, the Korean government also implemented the Knowledge Sharing Program (KSP) to pass on specific knowledge and skills developed in various public policy sectors to other developing countries. Transferring knowledge acquired in the development process is also a kind of self-regarding idea, in that it highlights what Korea can offer to other developing countries.

There are also other characteristics specific to Korean ODA in terms of the involvement of private actors. Compared to other donor countries, Korean ODA includes a large portion of concessionary loans in comparison to grants. Compared to 5.7% of the OECD/DAC countries on average, concessionary loans account for 38.3% of Korean ODA, while the grant portion accounts for the rest. Those concessionary loans are usually used for large-scale economic infrastructure such as airports, roads, and water supply facilities. Korean firms are often involved in these large infrastructure projects, a practice which is often criticized as advancing a donor country's own economic interests. Further empirical analysis shows that Korea provided more aid to higher–middle-income developing countries (E. M. Kim & Oh, 2012). The authors of the research argued that it indicated a tendency toward economic consideration in the practice of ODA in Korea.

It is also true that Korean people have a strong ethical obligation because they were given international relief and humanitarian aid in the 1950 and 1960s. It is worth remembering that after World War II, Korea received a large amount of international relief and development assistance. In particular, Korea received international aid during the Korean War (1950–1952), mainly from US and UN agencies such as the United Nations Korea Reconstructive Agency (UNKRA; Cha et al., 1997). The aid through UNKRA was about USD 120 million during the reconstruction period of 1953–1960, while aid from the USA reached USD 1,745 million, including Public Law 480 funds for food assistance (Cha et al., 1997: 13).

International support continued to come to Korea through the 1960s. It should be a strong obligation for those who have received aid from others in times of difficulties to return such support to other people in need. For Korea, it should be a strong ethical duty to provide international support for people in humanitarian crises and poverty. Ordinary citizens and civil society groups, indeed, volunteer to carry out humanitarian work. According to the National Survey on ODA in 2017, the first reason that Koreans support the provision of foreign aid is that Korea has also benefited from foreign aid in the past (Park & Kim, 2017). This rationale has always been listed as the first reason for supporting ODA, with higher priority than other rationales such as the universal purpose of international development for poverty reduction and global peace or the national economic and diplomatic interests of the nation. It is an important imperative to realize ordinary citizens' ethical obligation through government ODA policy, as this chapter attempts to do.

Reflecting on the experience of Korean ODA over the past two decades, an important underlying ethical rationale can be elicited. It can be summarized in the following ways. First, it is a significant symbol of national pride that Korea has become a donor country as a member of the OEDC/DAC (CIDC, 2017). Second, Korea has relevant experience and knowledge from which developing countries may learn for their advantage. Third, Korean ODA can also benefit Korean firms attempting to establish a business presence in developing countries. In a nutshell, Korean ODA is deeply based on self-regarding considerations. Such a self-regarding rationale may not necessarily be bad in and of itself as an idea behind ODA. It explains the inner motivation for ODA. Nevertheless, there is one critical element missing in this policy rationale: people in developing countries and their social and institutional challenges that Korean ODA is intended to support. The self-regarding rationale should be combined with the idea of ethical obligation in relation to ODA. The perspective of ODA policy also needs to take into account the dimension of global community, putting emphasis on creating the global public good. Fortunately, many Korean people feel an ethical obligation to help people in other countries because they have received support from the international community as a developing country. It is now necessary to reinforce the view that the Korean state should play its role as a moral agent that fulfills this moral obligation of its citizens. In other words, it should be made clear that the Korean state must provide help to other countries through ODA in the international community. In particular, it is important to construct the idea that there is an ethical obligation for Korea to undertake ODA beyond a self-regarding rationale.

5 Korean ODA Should Adopt a Three-Pronged Approach to Development

This chapter has raised the question of why Korea should support developing countries for their development. Entering the second decade as an OECD/DAC member, Korea needs a clear sense of purpose and a normative rationale for its ODA policy in order to contemplate future policy directions. The chapter has first discussed the normative theories that construct a moral duty and ethical obligation for international development assistance in order to set a normative reference. In particular, this chapter revealed that moral theories based on individual persons as moral agents cannot provide a sufficient imperative for ODA. Instead, it is necessary to regard the state as an ethical agent to provide support to developing countries for social and economic development. The chapter also argues that the state as a member of the global institutional order and an ethical agent responsible for and accountable to citizens needs to contribute to the creation of the global public good. Secondly, this chapter has reflected on the underlying policy rationale of Korean ODA. Behind the abstract ideas of human well-being and global prosperity that the Korean government officially states as goals of ODA, Korean ODA policy is strongly underlined by a self-regarding rationale of national pride and respect as well as economic considerations. Although Korean people have a very strong sense of ethical obligation since they received a large amount of international support in times of difficulties, Korean ODA policy needs to be firmly based on normative grounds in combination with the inner motivations of self-regard.

With such a normative discussion, this chapter suggests a three-pronged approach to Korean ODA for the future, while incorporating the strength of Korean development experience and knowledge: aid for human freedom, aid for socioeconomic and political development, and aid for the global public good. First, considering that human freedom is an intrinsic goal of development and that it is individual persons who actually suffer from poverty, disease, and violation, Korean ODA should aim to protect human freedom as a universal value, manifested in the Universal Declaration of Human Rights. The ODA policy of Korea should focus on eliminating the barriers to individual freedom in developing countries and establish this as the fundamental goal. For example, the central goal of ODA policy should be reducing the extreme poverty that still exists in many least-developed countries, liberating individuals from violence and conflicts in fragile countries, and applying a rights-based approach that promotes individual rights across a range of issues.

Korean ODA also needs to support developing countries to build effective institutions, which will provide options and areas to exercise freedom. This is an area in which Korea has strength. What is really important in implementing international development cooperation is establishing a self-reliant and sustainable basis for socioeconomic development in the future. Building effective public institutions means constructing an enabling environment. Further, Korean ODA should place a stronger emphasis on development effectiveness rather than aid effectiveness. Successful aid projects often do not contribute to the overall development of the country, which was described as a micro–macro paradox by Mosley (Mosley, 1986). The Korean government has succeeded in strengthening the evaluation of development projects over the past decade while also emphasizing aid effectiveness. However, there has been a lack of concern about development effectiveness that ultimately contributes to the overall development of partner countries. Korea's aid has grown to less than 2% of the total aid received by partner countries (KIEP, 2019), but the budget allocated to each project is too small and partitioned out to too many government organizations who want to engage in ODA. In this way, it cannot contribute to the socioeconomic development of the partner country. In order to achieve development effectiveness, it is necessary to introduce a program-based approach by increasing the overall size of the budget and extending the program implementation period. Moreover, aid should play a role as a catalyst that can lead to greater development effectiveness by engaging with the private sector. Further, in order to link the successes of microlevel projects to social development at the macrolevel, it is necessary for public institutions to work effectively. This is what we can learn from the Korean development experience, which is still relevant today.

Lastly, Korean ODA should also address challenges at the global level, such as climate change and air and sea pollution, so that the global public good is maintained. The global migration crisis and threats to global peace are also important challenges that require a strong response from Korean ODA. As the benefits will be evenly distributed when this global public good is provided, the Korean people will benefit, as will many other citizens in the world. Considering that Korea is one of the world's largest trading nations, Korea has a great responsibility for the provision of the global public good. To contribute to the global public good, Korea must pay attention to and actively engage in global issues. The most important global problems that will confront us within the next decade or so include climate change migration, terrorism, and conflict. However, the aid ratio specifically focused on climate change in Korean ODA was only 3% in 2014, while the DAC country average was 24% (OECD, 2016: 225). In addition, support for conflict and peace activities accounted for 0.3%, and support for internal refugees was 0.0% among total Korean ODA in 2017.

It is now time to move forward beyond sharing Korea's own development experience. Korean ODA needs to seriously address how to contribute to the development needs of partner countries and the global issues of the international community.[5]

References

Adelman, I. (1997). Social development in Korea, 1953–1993. In D. Cha, K. Kim, & D. Perkins (Eds.), *The Korean economy 1945–1995* (pp. 509–540). Seoul: Korea Development Institute.

Bellamy, A. J. (2011). *Global politics and the responsibility to protect: From words to deeds*. Routledge.

Bourdieu, P. (2004). *Science of science and reflexivity* (R. Nice, Ed.). Chicago: University of Chicago Press.

Center, & for Global Development. (2019). *Commitment to development index 2018*. Center for Global Development.

Cha, D., Kim, K. S., & Perkings, D. (Eds.). (1997). *The Korean economy 1945–1995: Performance and vision for the 21st century*. Seoul: Korea Development Institute.

CIDC. (2017). *2017 Korean ODA White Paper* (p. 181). Committee for International Development Cooperation.

Government of Korea. (2010). *The policy for advancing international development* (in Korean) (Office of the Prime Minister, Ed.). Government of Korea.

Government of Korea. (2015). *The second mid-term strategy of international development* (Office of the Prime Minister, Ed.). Government of Korea.

Hamati-Ataya, I. (2013). Reflectivity, reflexivity, reflexivism: IR's 'reflexive turn'—And beyond. *European Journal of International Relations, 19*(4), 669–694.

Harsanyi, J. (1982). Morality and the theory of rational behaviour. In A. Sen & B. Williams (Eds.), *Utilitarianism and beyond*. Cambridge: Cambridge University Press.

Hulme, D. (2015). *Global poverty: Global governance and poor people in the post-2015 era* (2nd ed., Global institutions, 100, pp. xxiv, 305). Routledge, Taylor & Francis Group.

Hulme, D. (2016). *Should rich nations help the poor?*. Cambridge University Press.

[5] www.stat.oecd.org, accessed on 5 November 2019.

Kabeer, N., & Cook, S. (2000). Editorial introduction: Re-visioning social policy in the south—Challenges and concepts. *Ids Bulletin-Institute of Development Studies, 31*(4), 1–10.

KIEP. (2019). *Country partner strategu mid-term review*. Sejong.

Kim, E. M., & Oh, J. (2012). Determinants of foreign aid: The case of South Korea. *Journal of East Asian Studies, 12*(2), 251–274.

Kim, S. (2011). Bridging troubled worlds? An analysis of the ethical case for South Korean aid. *Journal of International Development, 23*(6), 802–822.

Kohli, A. (2004). *State-directed development: Political power and industrialization in the global periphery*. Cambridge University Press.

Kwon, H. J., & Koo, M. G. (2014). *Korean government and public policies in a development nexus*. Cham: Springer, 2014.

Kwon, S. (1998). The Korean experience of poverty reduction: lessons and prospects. In KDI/UNDP (Ed.), *Poverty alleviation*. Korea Development Institute.

Lancaster, C. (2007). *Foreign aid: Diplomacy, development, domestic politics*. University of Chicago Press.

Midgley, J. (2014). *Social development: Theory and practice*. Sage.

Morgenthau, H. (1962). A political theory of foreign aid. *The American Political Science Review, 56*(2), 301–309.

Mosley, P. (1986). Aid-effectiveness: The micro-macro paradox. *IDS Bulletin, 17*(2), 22–27.

Nagel, T. (2005). The problem of global justice. *Philosophy & Public Affairs, 33*(2), 113–147.

OECD. (2008). *Development co-operation of the Republic of Korea: DAC special review*. OECD.

OECD. (2016). *Development co-operation report 2016*. OECD.

OECD. (2018). *OECD development co-operation peer reviews: Korea 2018* (p. 129). OECD.

OECD. (2019). Editorial: After the lockdown, a tightrope walk toward recovery. https://www.oecd-ilibrary.org/content/component/5f97c119-en. Accessed 19 November 2020 .

Park, J., & Kim, E. (2017). *2017 National survey on ODA*. Korea Institute of Public Administration.

Pogge, T. W. M. (2008). *World poverty and human rights: cosmopolitan responsibilities and reforms*. Polity.

Rawls, J. (1982). Social unity and primary goods. In A. Sen & B. Williams (Eds.), *Utilitarianism and beyond* (pp. 159–186). Cambridge University Press.

Ringen, S. (2007). *"Where does freedom come from?", What democracy is for*. Princeton University Press.

Sen, A. (1993). Capability and well-being. In M. C. Nussbaum and A. Sen (Eds.), *The quality of life* (p. 453). Oxford University Press.

Sen, A. (1999). *Development as freedom* (p. 366). New York: Alfred A. Knopf.
Singer, P. (1972). Famine, affluence, and morality. *Philosophy & Public Affairs, 1*(3), 229–243.
Song, B.-N. (1990). *The rise of the Korean economy*. Oxford University.
Stallings, B. (2017). *Promoting development: The political economy of East Asian foreign aid*. Springer Singapore.
United Nations. (2015). *The Millennium development goals report* (p. 72). UN.
Waltz, K. N. (1979). *Theory of international politics* (1st ed.). Boston, MA: McGraw-Hill).
Weiss, L. (1999). State power and the Asian crisis. *New Political Economy, 4*(3), 317–342.
Woo-, M. (Ed.). (1999). *The developmental state*. Cornell University Press.

CHAPTER 3

Pivotal Moments in Japanese ODA: Circa 1950–2010

Jin Sato

1 Key Determinants in the Evolution of Japan's ODA

What historical junctures determined the Japanese provision of foreign aid? While the history of Japanese institutional involvement in economic cooperation dates back to the early postwar period of the 1950s, key events that impacted the subsequent path of economic cooperation have

Japanese names here are written in Japanese order, with the family name first. This chapter is adapted from the forthcoming book of the same author Kaihatsu Kyoryoku no Tsukurare kata (The Make up of Development Cooperation) to be published from the University of Tokyo Press in May 2021.

J. Sato (✉)
Institute of Advanced Studies, University of Tokyo, Tokyo, Japan
e-mail: satoj@ioc.u-tokyo.ac.jp

H.-j. Kwon et al. (eds.), *International Development Cooperation of Japan and South Korea*,
https://doi.org/10.1007/978-981-16-4601-0_3

not been evenly distributed throughout history. There are some clusters of moments that shaped the subsequent path of aid institutions and practices.

This chapter highlights three of these clusters in turn: (1) how reparation was defined, and the administrative procedures for delivery of payments to claimants after World War II. The request-based principle (yōseishugi), a defining feature of Japan's ODA, was rooted in the reparations system; (2) the pressure from the Development Assistance Committee (DAC) and the United States to increase aid volume, which invited the Japanese government to use the Fiscal Investment and Loan Program (FILP). The rapid expansion of aid volume was instrumental to the Marcos scandal in the Philippines and the subsequent rise of criticism of Official Development Assistance (ODA) by the Japanese public in the 1980s and 1990s; and (3) the "discovery" and the incomplete intake of "human beings" as the major subject of development assistance facilitated by Ogata Sadako in the 2000s. "Human" here refers to the priority given to the "individual" over groups or the state in the process of development. Coming from a background in refugee assistance and peace building, Ogata had much skepticism toward the constructive role of the state. Instead, she emphasized the importance of "human" security which was rather alien to the work ethic of the Japanese and the basic assumption of ODA that relied on "Government to Government" agreements.

These pivotal moments reveal how current aid practices are historically grounded and how they were shaped by external forces, rather than intentionally derived and planned by the Japanese government. Understanding such external forces, therefore, is key to explaining Japan aid behavior. Let us examine the unique context of each of these "moments."

2 Stage One: Formation of a Mechanism for Resource Flow After World War II

2.1 *Economic Cooperation to Build Japan's own self-reliance*

What were the origins of Japan's foreign economic cooperation? Frequent references are made in Ministry of Foreign Affairs (MOFA) publications that the origins stem from "postwar reparations." Yet the question of how exactly reparations served as the origin of present-day development cooperation and aid often remains unanswered. Clarifying this issue should illuminate the historical trajectory of Japanese aid, from the economic cooperation of the 1950s to present-day development cooperation.

Why in the first place did Japan, struggling as it was in the early postwar years to get its own economy back on its feet, venture into Southeast Asia as a provider of economic cooperation? For several years after the end of World War II, Japan was far too destitute to give aid to other nations. The national income per capita in 1949 has been estimated as equivalent to US\$100—one-fourteenth of that of the United States that year and below that of nations such as Argentina, Mexico, and Turkey (Kindleberger, 1965: 12).

Japan's defeat in World War II gave her no choice but to provide reparations for the war damages. However, Japan had quietly worked out a strategy of economic cooperation *before* the conclusion of reparations agreements with Southeast Asian nations, and plans were moving ahead independently of its reparation obligations. To identify the real cause behind Japan's (re)entry to Southeast Asia it is useful to explain both its obligations and strategy in terms of their common factor: the domestic and international environment that fostered economic cooperation.

On the domestic front, there was an urgent need to tackle issues such as inflation and food shortages, but for Japan, which was under US control until September 1952, the United States' wishes loomed large and could not be ignored. After the Occupation ended in 1952, the most pressing task for Japan was to be accepted back into the international community, a goal that revolved around its efforts to obtain membership of the United Nations. For Japan, which had lost a major export market when China became a Communist nation, the only option was to gain recognition from developed Western nations by finding a new path in Southeast Asia and building up a track record in the field of aid.

The United States initially demanded punitive reparations that would not allow an industrial capacity greater than Japan's prewar level. In 1948, however, this punitive policy underwent a major turn aiming at building an independent Japan with a "reasonable standard of living." There was an about-face toward providing more aid to Japan, based on the view that an increase in its peacetime industrial capacity was desirable, mainly to use Japan as a bulwark against Communism with little additional burden on American taxpayers (Hamada, 1950). Based on the changing international situation in the region, such as the founding of the People's Republic of China in October 1949, there was a policy shift toward easing reparation demands and turning Japan into "Asia's factory" through democratization aimed at countering Communism. This change was of enormous significance to Japan because its economic independence relied on exports and imported raw materials, but a considerable portion of reparation negotiations between Japan and Southeast Asia involved the United States as intermediary (Kitaoka, 2015).

It was only natural that Japan would take an interest in the US vision of trilateral relations between the US, Japan, and Southeast Asia. This American concept aligned with the Japanese government's desire to compensate for the loss of the Chinese market, which after Communism's adoption was subject to embargoes, by developing the Southeast Asian market. If Japan could obtain raw materials from Southeast Asia and export to that same market, this would benefit Southeast Asian countries and reduce the foreign currency (US dollars) needed to purchase American products (Hays, 1971: 22).

Japan's first white paper on economic cooperation published in 1958, *Keizai kyōryoku no genjō to mondaiten* (Current situation and problems of economic cooperation), positioned the goals of economic cooperation at that time as outlined below, framing such cooperation as essential for the sake of the Japanese economy (MITI 1958: 22).

> Bolstering economic development in Third World countries promotes Japanese exports by stimulating those countries' demand for imports, and it can also help secure markets from which to import key raw materials. The future direction of Japan's industrial and trade structure particularly amplifies the need for this. In that sense, the dual aims of economic cooperation are like two sides of a coin and can proceed hand in hand.[1]

What is meant here by "Japan's industrial and trade structure" refers to an economic structure which fundamentally consists of (1) importation of raw materials, (2) domestic production, and (3) exporting of finished products. Whether in terms of seeking raw materials or seeking markets, resolving the reparations issue was crucial to reopening normal trade with Southeast Asian countries. The reparation method used at that time was to define the direction of Japanese aid for a long time hence.

[1] All translations are by the author.

2.2 The "Request-Based" Principle Evident in Reparations Procedures

In terms of methods of paying postwar reparations, the Treaty of Peace with Japan signed at San Francisco required that Japan make available "the services of the Japanese people in production, salvaging and other work for the Allied Powers in question," thereby stipulating that reparations take the form of "services." This policy was based on the lessons learned after the German economy was devastated following World War I because of the Allied Powers' demands for monetary and material reparations without considering defeated nations' ability to pay, which gave rise to radical forces such as the Nazis (Baishō Mondai Kenkyūkai [Reparations Research Group], 1963: 22). Regret for how the reparations methods imposed on the nations defeated in World War I brought about the Great Depression and World War II led to the requirement that Japan provide "services" rather than monetary and material reparations (Okano, 1958: 263). Notably, however, in actual negotiations with individual countries seeking reparations there was a gradual increase in demands for material reparations, and the scope of reparations broadened to include payment in the form of production goods. This was the origin of Japan's international yen loans.

Japan concluded successive reparations agreements withBurma (November 1954), the Philippines (May 1956), Indonesia (January 1958), and Vietnam (May 1959). Unlike earlier research, which has focused on the processes involved in this series of reparations negotiations, here I would like to focus on official procedures at the time of actual payment of reparations, after negotiations had been completed, because they formed the prototype for how subsequent development cooperation was implemented.

The first step was for each country seeking reparations to establish a permanent delegation in Tokyo to centralize their reparations affairs. The contact point on the Japanese side was the reparations section of MOFA's Asia Bureau. This unit served as the interface for negotiations with the Japanese government. Hayashi Yuichi, who was part of MOFA's reparations section, explained the general procedural sequence as follows (Hayashi, 1959: 13–14):

1. Each year the countries seeking reparations (a) make a list of the equipment and capital they wish to procure and consult with the Japanese side. This is called the annual implementation plan.
2. The upper limit of funds to be disbursed each year is determined separately for each nation seeking reparation, so in principle the estimate for capital goods such as construction materials is written into the implementation plan within that limit.
3. Once the implementation plan is agreed on with the Japanese government, the delegation from the country seeking reparations negotiates directly with Japanese businesses on that basis and enters into materials procurement contracts and services agreements. However, (b) when delegations do not know which company to choose, they often seek recommendations from the Japanese government (unlike with normal trade contracts, the Japanese government approves these contracts and payment of the contract money is made from the treasury).
4. Once the reparations agreement has been approved by the Japanese government and implemented, the government pays the contract money to a Japanese bank designated by the delegation, in line with the payment terms in the contract. This payment signifies that Japan has fulfilled its obligation to make reparations.

The underlined parts (a) and (b) in these procedures are the points to note here. Although the nations seeking reparations applied to the Japanese government for their desired items, Japanese companies would engage *in advance* in sales tactics aimed at the governments of those countries. Some apparently shared tips on how to obtain Japanese government approval or acted as proxies in preparing the reparation plans. This role of Japanese businesses in the formative stage of projects was inherited in subsequent loan projects (Ando, 1992: 31).

Why didn't Japan purchase capital goods directly from domestic companies and provide them as "Japanese products" to countries seeking reparation? The reason was that if the Japanese government does so, the recipient countries might face the risk of having to purchase poor-quality goods at high prices. In addition, with an indirect method that left negotiations over specific capital goods up to the parties concerned (the country seeking reparation and the company with which it had entered into a contractual agreement), the Japanese government could minimize administrative procedures for negotiating a price with the nation seeking reparation, and the entire transaction could be completed in yen without having to pay out foreign currency (Baishō Mondai Kenkyūkai, 1963: 100).[2] Just as regrets over the aftermath of World War I reparations demanded of Germany, which in principle required monetary payments, had a major effect on subsequent Japanese reparation methods, the fact that Japanese reparations were restricted to goods and services was a strong determinant of later methods of providing aid and economic cooperation.

It should be noted, however, that Japan's emphasis on capital goods as the foundation of World War II reparations was not solely because of Japan's tight financial situation. There was the idea that one aspect of reparations is atonement, so to improve recipient countries' attitudes toward Japan, capital goods—which provide a long-lasting impression of reparations—were preferable to consumable goods that soon disappear (Baishō Mondai Kenkyūkai 1963: 96). The construction of infrastructure such as dams and power plants, which would later come to symbolize Japanese ODA, also received support on the grounds that publicity could be expected through the method of reparation and the infrastructure's "afterimage." With its name changed to the "request-based principle," this mechanism from the reparations period was handed on to the next phase of development assistance without fostering a policy culture of ascertaining the outcome of economic cooperation on the ground. The decision not to hammer out a philosophy of economic cooperation can be regarded as one explanation for the lack of a foundational principle of Japanese aid, which has been criticized as "faceless."

2.3 Formation of a System to Promote Economic Cooperation

The creation of an economic cooperation system in Japan was also rapidly accelerating in response to the dramatically changing international environment. In June 1953, different nerve centers for economic cooperation policy were established in quick succession, including MOFA's Asian Economy Discussion Group and the Asian Economic Cooperation Committee of the then-Liberal Party. In particular, the Asian Economy Discussion Group (*Ajia Keizai Kondankai*) under MOFA's auspices was

[2] The exchange rate at that time was fixed at 360 yen to the dollar, so there was no need for concerns over currency fluctuations, but there was the problem of a steep rise in the price of Japanese products (Okano, 1958: 568).

set up in June 1953 for representatives of private-sector firms and high-ranking government officials, including the foreign affairs minister, to discuss economic cooperation to Asia from a broad perspective and to determine a basic policy. The fact that influential figures from different sectors, including the foreign affairs minister, met weekly vividly illustrates just how important economic cooperation was to Japan at that time. The Asian Economy Discussion Group gave birth to the Society for Economic Cooperation in Asia (which later became the Overseas Technical Cooperation Agency and then JICA). This acted as the first comprehensive postwar private aid organization, and it played a key role in formulating the basic framework for postwar Japan's overseas economic cooperation.

These proactive moves by the Japanese government stemmed from the international environment at that time. The government, seeking a reduction in discriminatory measures such as credit limits and tariff barriers that hindered free trade by Japan, feared being left behind by Western countries, which were moving ahead with the formation of economic blocs. In this situation, Japan had hoped that economic cooperation would serve as a way to achieve an international balance of payments that did not rely on special procurements and US aid to Japan (Akiyama, 1961), both of which kept Japan in the shadow of the US.

The Cabinet decision on economic cooperation for Asian nations that was reached on December 18, 1953 was a key declaration plainly indicating that Japanese economic cooperation kicked off under private-sector leadership. This decision clearly set forth a policy whereby economic cooperation was in principle to be carried out through private-sector initiatives, with the government providing the assistance necessary for its implementation. The idea of positioning the private sector as the key player in economic cooperation was not, however, the outcome of an autonomous and strategic decision. The situation at the time was such that this was the only way to avoid giving the impression that economic cooperation constituted an economic invasion by the Japanese government. Ultimately, rather than being a matter of the government failing to take the lead or choosing not to take the lead, it was a matter of not being able to do so, given the state of international affairs and government finances at the time.[3]

[3] The reason that Japan frequently used the term *keizai kyōryoku* (economic cooperation) rather than the term *foreign aid* that was widely used in the West was not only that Japanese aid was motivated by business, such as trade and investments, but also because

Although economic cooperation was closely related to reparations, it was conceived of as an entirely separate approach with its own strategic aims. It clearly existed as a private-sector-driven means of promoting exports and securing raw materials, quite apart from reparations, which emphasized government-led provision of services. Resolving the issue of reparations was a facilitator to the promotion of economic cooperation in the broad sense. Japan created its system of economic cooperation in a back-and-forth between an abstract discourse on philosophy and policy and a concrete discourse on implementation procedures.

3 Stage II: From Economic Cooperation to Development Aid

3.1 The Expansion of ODA

Even after the high economic growth of the 1960s had ended, the Japanese economy continued to expand. The gross domestic product (GDP) witnessed a stunning five-fold increase from 1971 to 1990 and, as Fig. 1 shows, the ODA budget for this same period grew around 18-fold (Imazu, 2009: 71).

How was Japan able to boost its overseas aid so rapidly from the 1970s into the 1980s despite facing domestic problems such as social infrastructure including a sewage system. Even if the rapid appreciation of the yen undoubtedly boosted Japan's ODA work, we must look elsewhere for the reason the government could treat foreign aid as "sacred."[4]

Let us first note the circumstances during that period. The mid-sixties were a pivotal time when Japanese economic cooperation shifted to Western-style "development assistance." Having successfully hosted the 1964 Olympics, Japan joined the OECD that same year, becoming the first Asian nation to do so. It initiated large-scale yen loans to South Korea based on the 1965 Treaty on Basic Relations between Japan and the *Republic of Korea, and in the following year it played a key role in the establishment of the Asian Development Bank*. Underpinned by striking

it was necessary to choose a term that could be readily used in relation to Asian nations wary of new Japanese advances into the region.

[4] The value of the yen, which stood at around 240 yen to the US dollar in 1985, rose to 150 yen in a mere two years.

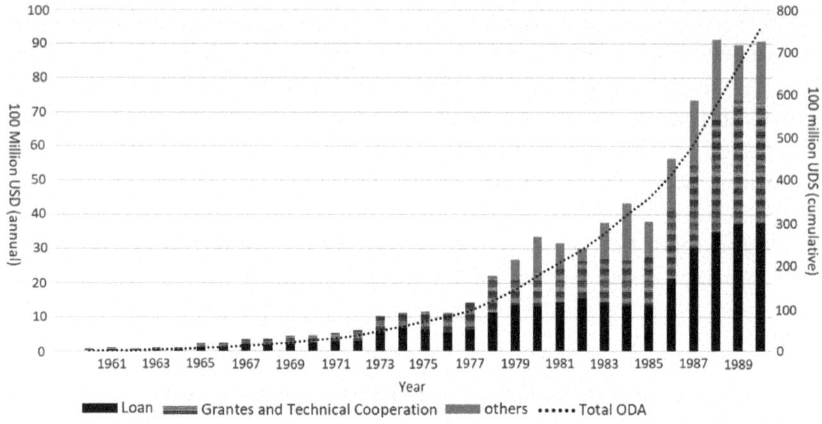

Fig. 1 Breakdown of Japanese ODA by year, and the cumulative total of ODA (*Note* The figure for "Loans" represents the remainder when ODA loan repayments are deducted from ODA gross loans. The figures for "Grants" and "Technical cooperation" are taken from the "Grants, Total" item. *Source* By the author from https://stats.oecd.org/qwids/ for ODA and from World Development Indicators for the GDP growth rate)

economic growth in the 1960s, Japan carved out an international position as an aid provider, and the world had expectations that it would increase its aid.

The United Nations Decade of Development that President Kennedy launched in 1961 ended up as a "Decade of Disappointment," and it was not possible to achieve the initial target of a 5% economic growth rate in all developing countries and an allocation of 1% of advanced industrial nations' GNP to aid to achieve that target. Not only did the gap between North and South widen, but the inequities within developing countries also accelerated (Tanaka, 2003).

The growing international criticism and pressure in relation to Japanese aid in the 1970s were an extension of this trend. Japan's economic status was now such that its excuse to the West that it could not expand ODA because it was a late starter as an aid provider no longer held water. The trade imbalance and Japanese products, which had rapidly penetrated these markets, galvanized local pride, leading to anti-Japanese movements around Southeast Asia (Pattajit, 2017). Japan was exposed to waves of pressure and demands at home and abroad, and there was no leeway to formulate policies independently and proactively. What, then, was the nature of the pressures that characterized Japan's ODA policy? Below I focus on DAC and the United States and examine the role that each actor played in expanding aid.

3.2 *Pressures from DAC and the US*

In the initial stage, DAC's criticism of Japanwas directed toward its low level of aid, but from the 1980s the criticism focused on the quality of the aid. An important function of DAC is to keep Western aid providers in alignment, and the relative lag on Japan's part gradually became noticeable, including in its compilation of statistics. An analysis of DAC minutes for the past three decades reveals that from the 1980s into the 1990s demands for a quantitative increase in aid from Japan subsided and the focus shifted to quality (Maemura, 2019). In particular, DAC was a critique of the low grant elements and of the "tied" condition.

"Tied" assistance limits equipment suppliers to Japanese companies.[5] This has the effect of restricting free choice in the international market, which from recipient nations' perspective forces them to purchase goods at relatively high prices. The calculations used by DAC agreed with the recipient nations' perspective. When recipient nations are not only unable to choose the goods they need but are also obliged to purchase from abroad even those goods that can be procured domestically, this might hold back domestic industries (Prime Minister's Secretariat, 1970). The West and aid recipient nations framed Japanese aid as nothing more than a brazen pursuit of national interest in the name of "aid." It contained too few gift elements.

Let us trace how the brunt of criticism shifted from quantity to quality, based on DAC reviews of Japan in 1962, 1968, and 1973. In 1962, shortly after Japan became a member of DAC, Japan was chosen for a DAC review and came in for criticism in terms of how it compared internationally. In the investigations at that time, Western nations peppered the Japanese government with questions on 15 items, such as the purpose of

[5] A grant element is an indicator of the concessionality (softness) of aid conditions. If we posit commercial loans as grant element 0%, the figure rises as conditions (interest rate, maturity, grace period) are eased; with gifts it stands at 100%. To qualify as ODA, loans must incorporate a grant element of at least 25% (*Kokusai kyōryoku yōgoshū* [Lexicon of international cooperation], 3rd ed., 59).

aid and the government's views on tied aid, as well as the outlook for private capital (DAC, 1962). The fact that questions focused on the state of aid shows that advanced Western nations at that time simply wanted to know the circumstances surrounding aid from Japan, which was a new aid provider. We can also sense them turning a temporary blind eye to the low level of gifts in Japanese aid, based on the rationale that per capita income in Japan was lower than Western aid providers.

In the 1968 review of Japan, when it was starting to compare favorably with the West as an aid provider, the criticisms were harsher. Commenting on Japanese ODA, the Canadian representative stated that if providing aid was inherently regarded as an act of self-sacrifice, then Japan's vigorous advocacy of fair burden-sharing based on per capita national income was somewhat dubious. The Japanese representative from MOFA rebutted this criticism, arguing that it was impossible to fairly evaluate the aid efforts of an underdeveloped nation with rapid economic growth such as Japan based solely on national income and GNP (Ministry of Foreign Affairs, 1968: 18).

In the 1973 review of Japan, the US representative declared that Japan was the only major aid provider to depart so greatly from DAC's recommendations on conditions and that Japan's most urgent task was to ease aid conditions, and criticisms of Japan became more vociferous (Ministry of Foreign Affairs, 1973: 26). Other nations that submitted comments for the review—in relation to "tied aid," for instance—did commend Japan to some extent for its efforts to ease procurement conditions. Nevertheless, they were critical of the lack of clarity as to when and to which nations and under what circumstances the conditions would be eased (Ministry of Foreign Affairs, 1973). The MOFA report also records that these critics regarded the domestic setup, such as public understanding and the personnel shortage, as the reason that Japan could not become an aid provider that met Western expectations.[6]

Perhaps the greatest pressure on Japan to boost the amount of aid was from the US government. For instance, it is well known that in 1965 the US pressured Japan to support the new Suharto regime, which was leaning toward the Western camp. U. Alexis *Johnson, the then US*

[6] At this meeting, Japan stated that it completely agreed on the need for more staff but that it was government policy to reduce the quota of personnel in government offices as much as possible, so a personnel increase would be difficult (Ministry of Foreign Affairs, 1973: 18).

ambassador to Japan, was critical of how Japan's investment in the Asian Development Bank, which was to be established in 1966, fell far below US expectations, and he recalls that "The negotiations over Ogasawara and Ryūkyū had somewhat of an effect" (Johnson 1989: 170–71). This reveals that Japan's Southeast Asian involvement was intertwined with domestic issues concerning the reversion of control over the Ogasawara Islands and Okinawa to Japan and that this was a factor in negotiations with the United States.

Where pressure from the United States was most vividly reflected was in the easing of conditions on tied loans. Figure 2 shows how the tied ratio rapidly declined in the 1980s, when US pressure increased. This was

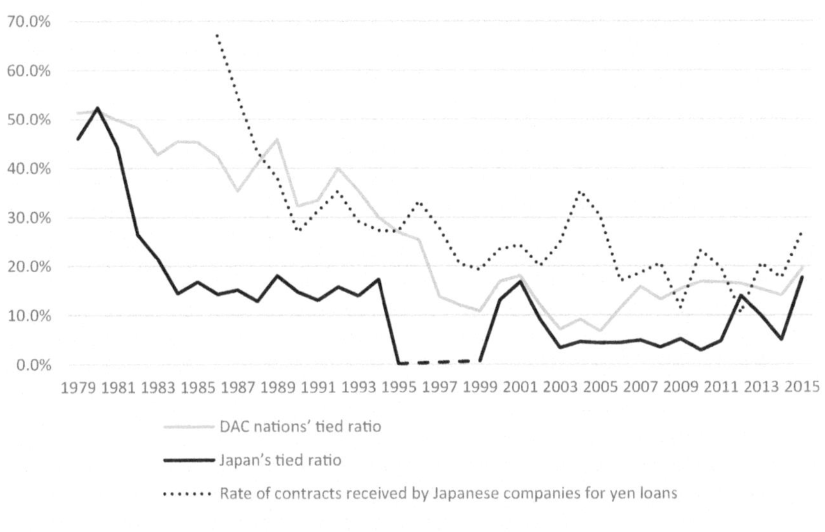

Fig. 2 The ratio of tied yen loans and the rate of contracts received by Japanese Companies (*Note* Although the operational part of Japan's gratis fund aid is tied to the main Japanese parties to the contract, the reason the tied ratio is sometimes close to zero is that procurement of material and equipment, services, and labor from third countries is not excluded. Hence MOFA regards grant aid as untied and has reported as such to DAC. Nevertheless, approval from the Japanese side is necessary when Japan procures these items from a third country, so DAC has regarded Japan's grant aid as "tied" and has criticized this practice (Hamana, 2017: 188). *Sources* Annual reports of OECF, JBIC, JICA)

in response to the rise in external pressure in reaction to Japan's growing balance of payments surplus, and in 1978 it was initially decided to boost the untied ratio of aid to facilitate the entry of US companies in particular. This trend toward overall easing of tied conditions subsequently accelerated (Lancaster, 2010: 39).

These facts were not, however, widely known among foreign companies that had the opportunity to submit bids. In particular, there were no signs of these facts resonating in the United States (Fujimura, 1992). In fact, as shown in Fig. 2, even in FY1997, when the tied ratio for Japanese businesses' yen loan contracts was almost zero, it was possible to maintain contracts at around the 30% mark. In the first place, most ODA from Europe and the United States when Japan was becoming a major aid power consisted of gratis fund aid, so it is not possible to make a simplistic comparison of their tied ratios with that of Japan, where yen loans accounted for the major portion. Nevertheless, we should note that the tied ratio of DAC member countries, including that of the US, did not fall as much as Japan's, despite DAC criticisms of tied aid.

Japan had to respond by expanding the ODA budgetwhich was almost impossible just by relying on general account budget which was strictly monitored by the Ministry of Finance. A new opportunity arose when OECF (Overseas Economic Cooperation Fund) found Fiscal Investment and Loan Program (FILP) as source of their budget. It is the state owned fund through government financial institutions (most notably postal savings) to finance public projects undertaken by government-affiliated corporations or to finance government loans to borrowers in targeted areas (targeted industries, small firms, mortgage borrowers, etc.). It was through this scheme that Japan was able to mobilize yen loans to expand ODA in a dramatic way and fund infrastructure abroad as if they are the extension of its domestic projects.[7]

In this way, although foreign criticismsthat Japan's aid was low were aimed directly at MOFA, Ministry of Trade and Industry (MITI)—which has great influence over domestic industrial policy—was able to continue Japanese-style economic cooperation, regarding it as "an extension of domestic public works" while keeping an eye on criticisms from abroad. Compared with domestic policies, where the legislature participates in the

[7] In 1991 when Japan was among the top ODA donor, 52% of the ODA budget was from general budget account while 43% came from FILP. FILP thus had the effect of reducing the budget constrain and the pressure to increase tax (Takehara, 2014: 146).

policy-making process and it is easy to incorporate criticisms into policies, the structure of decision-making in foreign policy, where the administration (i.e., MOFA) apparently had broad discretionary power, was also associated with Japan's response to international criticisms (Kusano, 1989: 56).

In other words, it is likely that the reason that pressure does not result in radical structural reforms is the weak link between such pressure and domestic stakeholders. Even if there were quantitative changes, such as in the tied ratio, what undermined the aid system itself was the upsurge in *domestic* criticism, as discussed below.

3.3 The Marcos Scandal and Criticisms of aid

In 1985 the Marcos regime's corruption in connection with ODA was revealed with clear evidence indicating behind-the-scenes connections between Japanese companies and the Filipino regime. As a result of this scandal, there was a sudden upsurge in domestic criticism of ODA, which greatly affected the public's views on aid. The fact that Marcos had cannily diverted foreign funds to line his own pockets emerged through rigorous investigations instigated by President Corazon Aquino, who took power after Marcos' downfall. According to one source, the wealth that Marcos illegally amassed during his presidency totaled somewhere between 11.6 and 22.3 billion dollars (Hunt, 2013).

Japanese ODA must have been very convenient for Marcos, based as it was on the request-based principle. Unlike in the United States, where the donor decides on the projects, the request-based principle gave aid recipients room for flexible negotiation. In particular, product loans were provided to help cover shortages in the foreign currency needed for developing countries' imports. These loans were not tied to specific projects, the procedures were straightforward, and the loans could be approved quickly.[8] In addition, the Japanese non-imposition of any conditionality (aid conditions) whatsoever, including no negotiations over the schedule for discharging the debt, effectively functioned to give Marcos a free hand (Rivera, 2003: 525).

How did this arrangement come about? According to a critical biography by Cesar Virata, who as finance minister long served as Marcos'

[8] That is precisely why product loans were criticized as a breeding ground for lining the pockets of corrupt regimes (Sumi, 1989: 19).

right-hand man, Japanese yen loans—of which product loans constituted the bulk—were superior in terms of the speed of the procedures necessary before the loan could be used (Sicat, 2014: 285). Amidst the decline in US aid to the Philippines in the 1980s, yen loans, whose lending criteria were less stringent than those of institutions such as the World Bank and which were on a larger scale, made deep inroads into the Marcos regime. It is thought that the mechanism of direct negotiations between recipient governments and Japanese businesses that commenced in the reparations period discussed above-created conditions where it was easy to collect a "handling fee" from businesses seeking a contract for a project.[9] At the end of the 1980s, criticism of aid reached a peak in Japan. Various newspaper articles, as well as publications aimed at the public, discussed problems with ODA and even cast doubt on its very rationale. The rise of ODA criticism in Japan galvanized sectors of the public that had until then shown virtually no interest in development cooperation, and it contributed to a general awareness of ODA as an important policy area.

Some individual criticisms of ODA lacked an adequate basis in fieldwork. Yet the appearance of entities that took on the task of fieldwork-based communication in the area of development cooperation, whose information sources had until then been severely constrained because of the reliance on government-controlled information, marked the advent of a new era. Until then information on ODA had been available in grey documents by MOFA and MITI but was not readily accessible to the public.

Critiques of ODA, a headache for the government, still played a constructive role educating the media and the populace, beyond the stakeholders directly involved. Triggered by this groundswell of criticism, ODA also began to attract attention from members of the Japanese Diet (parliament), and a monitoring system was put in place. The process of opposition—for example, failed movements again dam construction—roused a sense of rights among people and gave them the opportunity to learn how to work in solidarity with external organizations.

[9] Mendoza (2001: 47) states that unless private businesses paid this fee, they could not receive contracts for projects.

4 Stage III: Toward People-Oriented Development, But Not Quite

4.1 *The "Discovery of Human" in ODA*

The end of the Cold War greatly altered the significance of development aid. The US and the Soviet Union, the two major world powers, no longer competed through aid to the Third World, and international society began to focus more directly on the public good, such as environmental conservation and poverty reduction. The publication of the Human Development Report by the United Nations Development Programme (UNDP) symbolized this change. Along with new indicators, there was a growing awareness of and emphasis on the fact that what was important was not GDP growth, but people-oriented development.

Referring to examples such as Isaac Newton, who explained how an apple falls, Mahbub ul Haq, a Pakistani economist who contributed to a major shift in the international development discourse toward a focus on human beings, stated that "the most difficult thing in life is to discover the obvious" (Huq, 1995: 3). In Japan, where the overseas activities of NGOs and charity organizations got off to a later start than in Western countries, development cooperation stemmed from reparations, whereby aid moved forward under government leadership rather than through a focus on people. In the world of development cooperation, which has been governed by numbers, including GDP, it took a long time to discover the 'obvious' fact that it is human beings who are both the means and the goal of development.

It was in the latter half of the 1990s that the perspective of people-oriented development was incorporated in Japanese development cooperation in earnest. In May 1998, then-Minister of Foreign Affairs Obuchi Keizō (later prime minister) first presented the concept of "human security" in a policy speech in Singapore, and this rapidly took off as an aid ideal with broad applicability (Yanagihara, 2019: 7). The consistent emphasis here was human security as a concept that saw human beings as individuals and aimed to ensure their freedom and potential (Ueda, 2010: 8). In a public relations pamphlet, MOFA stated that Japan was striving to make the twenty-first century a human-centered century (Ministry of Foreign Affairs, 2007: 4).

In this way, aid entered a phase in which it must be justified in the light of universal values unrelated to the ideological conflict between East and West—values such as environmental conservation, poverty reduction, refugee protection, and peace. The concern for the human condition, which during the Cold War had been overshadowed by national issues, finally came to light. Moreover, it was obvious that these development issues were too broad for governments to manage alone. For Japan, which needed wide-ranging public support and participation to confront these issues, it was no longer acceptable to not have a proactive philosophy.

4.2 The Four Guidelines and the ODA Charter

Japanese attitudes toward ODA after the Cold War were greatly affected by the Myanmar pro-democracy movement led by Aung San Suu Kyi in 1988, the Tiananmen Square protests (June 4, 1989), Iraq's invasion of Kuwait in August 1990 (the Gulf War), and international society's reactions to these events. With the invasion of Kuwait, Japan could not dispatch personnel to join the increasingly unified US-led Western forces, but it contributed 13.5 billion dollars to help cover the Coalition's costs. Nevertheless, it was not included in the Kuwaiti government's letter of gratitude, and the United States criticized Japan's contribution as "too little, too late." This apparently led to a strong sense of defeat not only within the ruling party but also in MOFA (Shimomura et al., 1999: 81).

What was presented as a response to these international situations and the mounting domestic criticism of Japan's lack of an aid philosophy was the Four Guidelines of ODA suggested by then-Prime Minister Kaifu Toshiki. These Guidelines constituted a philosophy of making comprehensive decisions about bilateral relations while paying attention to the following four factors when distributing ODA: (1) trends in military expenditures of recipients, (2) trends in the development and production of weapons of mass destruction, (3) trends in the export and import of arms, and (4) efforts to promote democratization, the introduction of a market-oriented economy, and the securing of basic human rights and freedoms.

In June 1992, environmental conservation and a ban on utilizing aid for military purposes were added to these Four Guidelines and the whole document was made public as the ODA Charter, an initiative of the Miyazawa cabinet. Japan had long adopted a principle of non-interference in recipient nations' domestic affairs, and the request-based principle had been a feature of its aid, but the end of the Cold War and the fact that Western donors no longer needed to worry about losing developing nations' support lay behind Japan taking a further step and acquiring political conditionality as a means (Shimomura et al., 1999: 5).

In ODA history from the 1990s onward, there were multiple occasions that put the implementation of Japan's philosophy to the test. Of special note is the suspension of gratis fund aid toChina in 1995. Also noteworthy is the fact that, unlike the suspension of aid that Japan implemented as a "sanction" for the 1989 Tiananmen Square incident in line with the West, the 1995 suspension of gratis fund aid to China was based on independent Japanese policy (Katada, 2001). In other words, this suspension did not occur passively as the outcome of acting in concert with the West, as had been the case in the past; instead, it was a litmus test showing that Japan could autonomously apply its own philosophy even in relation to China, with which it has been economically interdependent.

4.3 Ogata Reforms and "Human Security"

A major catalyst for Japanese aid principles taking on a form that could appeal to international society was the appointment of Sadako Ogata as JICA president (2003–2013). Sought after for her rich international experience, Ogata moved to JICA on October 1, 2003 and took up a position as its first president coming from the non-governmental background, striving to put Japanese aid principles into practice. Having served for many years as the United Nations High Commissioner for Refugees (UNHCR), Ogata had witnessed far too many situations where nations in a position to protect people were engaged in crushing oppression. Ogata's idea of changing the nation-based approach to ODA and moving to a focus on individuals was underpinned by her on-the-ground experiences as UN High Commissioner. From the perspective of Ogata, "... not only were nations unable to protect people, but the nations were themselves part of the problem" (Nobayashi & Naya, 2015: 227).

MOFA defines human securityas a concept that promotes sustainable individual autonomy and the creation of sustainable society through protection and empowerment, focusing on individual human beings in order to protect them from wide-ranging and serious threats to their lives, livelihoods, and dignity and to allow them to realize the rich potential they each possess (MOFA, 2020).

To implement this principle, Ogata particularly emphasized listening to frontline workers and empowering them by handing to them the power to make decisions. Concrete measures involved a major transfer of personnel from JICA's Tokyo headquarters to overseas offices and a transfer of authority.[10] By the end of FY2006, the number of personnel in overseas offices had increased by about 20% (195 people), and headquarters personnel and overseas personnel were more or less equal in number (JICA, 2019: 32). This resulted in concrete improvements, such as a major reduction in the time lag between applying for ODA and receiving it, something that had been criticized as frustratingly slow. One major change I noticed when visiting one of these local offices was the uptick in hiring of local staff. Noteworthy were progressive moves such as hiring local people not only for low-level positions such as drivers and support staff but also at the level of assistant director of local offices. Initiatives such as moving away from the previous emphasis on Asia and bolstering operations in Africa and strengthening support for peace-building symbolize the changes in development cooperation during this period.

Peace-buildingwas clearly positioned in terms of continuity with development, and the previously separate concept of "human security" was now added to that of "people-oriented development," which had already become widespread.[11] Nevertheless, we cannot ignore the fact that the concept of human security, originally introduced as a counter to national security, was ultimately introduced as an add-on to each nation's sovereignty. The UN Resolution on Human Security that was adopted on September 10, 2012 confirmed "full respect for the sovereignty of States, territorial integrity and non-interference in matters that are essentially within the domestic jurisdiction of States," positioning sovereignty as the superordinate concept (Ministry of Foreign Affairs, 2012).

Hence the discourse on human security generally failed to regulate realities on the ground (Yanagihara, 2019). Commenting on the series of reforms implemented by Ogata, the official organizational history *Kikō-shi* (The history of JICA) bluntly states that the staff were confused, although it does not go into details (JICA, 2019: 31). Specifically, there was a discrepancy between Ogata's ideals, which sought to incorporate human

[10] In addition, new regional support offices were established at six sites to strengthen on-the-ground operations around the world and each office's activities, and they are effectively tackling shared challenges in each region. JICA's domestic headquarters was also reorganized, and steps were taken to simplify and expedite decision-making processes and to make Japanese support for on-the-ground operations more efficient.

[11] The Human Development Division was established in JICA in FY2004.

security based on her experiences with refugees and peace-building, and the realities of JICA, whose work had until then focused on development (interview with JICA leadership, July 30, 2020).

Mainstreaming human security was a difficult challenge even for Ogata, who had decisively implemented various bottom-up approaches. Yet that does not signal the futility of a focus on people. For Japanese staff at JICA, who were mostly used to regarding people as part of a collective, the concept of emphasizing the individual was theoretically understandable, but nevertheless difficult to accommodate in their work culture. Rather than seeking a solution midway between the individual and the collective, we must pursue a philosophy aligned with conditions on the ground, functioning within the tension among diverse values. The more the discourse leans toward nation-building, the more essential is a concept that pulls it back toward the opposite pole, to stabilize the common ground somewhere in the middle.

4.4 Foregrounding National Interests

Ironically, the time when Ogata was trying to focus more on people than nation-states and was attempting a radical revamp of ODA was also a time of vigorous debate over ODA's contribution to national interests. In the final report of the second ODA reform panel in 2002, which was the forerunner of revisions to the ODA Charter in 2003, a statement that using ODA effectively as a means of foreign diplomacy would remain important for Japanese interests in the future was accompanied by an emphasis on the fact that securing ODA was crucial for Japan's own sake. It is unlikely, however, that members of a colloquium convened by MOFA and prioritizing foreign relations were envisaging an ODA that openly pursued domestic interests. Rather, in a situation of declining finances it is quite likely that they chose to emphasize national interests as a means of securing an ODA budget closer to their desired size. There was particular dissatisfaction with ODA among business circles, which were suffering from an economic slump. The Japan Business Federation "Opinion on the review of the ODA Charter" (Keidanren, 2003) stated that the stance of proactively implementing ODA should be clearly set forth in the Charter to secure the safety and prosperity of Japan as a trading nation poor in domestic resources—a statement that ventured into the pursuit of national interests from a protectionist private-sector perspective.

Obviously, the concept of national interest already existed in Japanese foreign policy (Ōyama, 2018). At issue here is whether this existed below the surface—lingering quietly as a tacit understanding—or whether it surfaced as an explicit argument. Even more important was the question of what forces were at work to bring this to the surface. From around 2003, when there arose an active debate in Japan on whether to cease aid to China, given its striking economic growth and military expenditure, this concept made an appearance when the ODA Charter was being revised (2003). A report titled "Open promotion of national interests" appeared as part of the ODA review carried out under the Democratic Party government's Foreign Minister Okada Katsuya in 2010 (Ministry of Foreign Affairs, 2010).

From the perspective of a historical genealogy of development cooperation, we can conclude that the idea of "open national interests" that was repeatedly highlighted in the 2015 revision of the Charter was merely "foregrounded" (Ōyama, 2019). Development cooperation is carried out through taxes collected more or less compulsorily from taxpayers. Looked at in this light, ODA was essentially a device for promoting national interests.

4.5 *The Culture of Doing Development*

How has Japan viewed the people who are the ultimate recipients in the context of development cooperation? The philosophy of self-help and autonomy was introduced to Japan in the early Meiji period (1868–1912) alongside Western European culture in the form of individual autonomy. Under the favorable conditions of a historical turning point in the form of the Meiji Restoration, the concepts of individual autonomy and self-help were met with great enthusiasm, yet they were not powerful enough to penetrate the depths of Japanese culture. The tradition of collectivism stood as a barrier at all levels of society. Hence it was no anomaly that at some point the discourse of self-help, originally imported with the "individual" in mind, replaced the individual subject with "autonomy of the nation and region" and was transplanted into the context of development cooperation.

One reason why we do not notice this shift is the obfuscation of subject in the widely used expression "support for self-help efforts." It is unclear as to exactly who should be self-supporting (Udagawa, 2017: 122). The implicit subject of the self-efforts that Japan touts is the nation, the region, the village; rarely was the individual envisaged in this context. It was this history that lay behind Ogata's eventual efforts to bring the focus back to "human beings" under the philosophy of "human security." The ambiguity of the subject of autonomy and self-help made the concept of self-help efforts seem convenient, but applying this philosophy in the field exposed the discrepancy between the nation (an institution) and the people (individuals).

Excessive collectivism risks entrenching the inequities and inequalities present within a group. That is precisely why Japan has regularly needed individualism. In a society that regards selflessness as a virtue, a society where it has become the norm to consider sacrifice of the few as inevitable, we also need to harken to the SDG pledge to "leave no one behind." Japan has wavered back and forth amidst the tension between the collective and the individual. This vacillating is also reflected in views on development cooperation.

Japanese collectivism fits well with industrialization, and it was useful for promoting Japan's own modernization and development. Collective job-hunting (i.e., recruiting young workers en masse in the countryside and forcing them to move to big cities) during the period of high economic growth was a classic example of this. In the initial stages of modernization and industrialization, the focus was on the oppressive and restrictive aspects of the traditional nation-state and local communities, and the emphasis was on liberating individuals from such confining circumstances. Yet the relationship between individuals and the collective had been cultivated over a long history and was not something that could change overnight. Even for individuals who had left traditional rural society and moved to a city, the culture that emphasized the collective continued in different forms, such as the collective constituted by a company.

Whether Japan's development cooperation contributed to recipient nations' autonomy is an important question. Nevertheless, to make this question meaningful we need to clarify the subject of autonomy in any given context. National autonomy and individual autonomy operate in different dimensions, and the two are not necessarily compatible. What is important to Japan is to sum up the strengths and weaknesses of collectivism and verbalize these in readily comprehensible language as a philosophy. In today's world, where society based on competition and individualism is showing signs of having reached an impasse, this also means that Japan and other Asian nations need to demonstrate an alternative kind of development and present the possibility of a multi-polar world.

5 Conclusion

The Japanese government accepted civil society's growing criticisms of ODA from the late 1980s into the 1990s as an unforeseen event and made intentional efforts to incorporate these criticisms into its subsequent operations. Many specific measures proposed by the bestselling book *ODA enjo no genjitsu* (The realities of ODA) by Sumi Kazuo (1989), which epitomizes criticism of ODA—such as considering the environmental impact, improving the research setup, and making information publicly available—were all institutionalized in some way by 2020, during the three decades since its publication. Nevertheless, improving the system and facilities for development cooperation is not the same as eradicating development challenges.

More than efforts to improve individual projects, what is now needed is to nurture a media environment and a populace that will critically oversee the work of development cooperation from a long-term perspective. Just as projects in the past were forged in various ways through criticism, development cooperation in the future also needs opportunities to relativize itself through outside voices. A range of measures can be implemented, from major goals such as training journalists who specialize in covering aid and providing support so that local people on the ground where aid operations are carried out can themselves disseminate information, to easy-to-implement measures such as actively introducing debates into university classes on international cooperation.

If we assume that development cooperation sets out from negotiations around the needs of both the donor and recipient and that other needs arise even after these have been met, then the perfect development operation that satisfies everyone does not exist. In that case, constructive criticism should always be necessary to keep development cooperation meaningful to the people in the area where the aid is provided. Historically speaking, development cooperation has undoubtedly strengthened the power of nation-states. In that case, nurturing in developing nations a civil society capable of critiquing nation-states' development operations has been an ongoing task, and remains a critical project.

References

Akiyama, A. (1961, September). Wagakuni no "Enjo" Seisaku no Shindankai [The new phase of Foreign Aid in our country] *Keizai Hyoron*, pp. 46–55. (in Japanese).

Ando, M. (1992). *Seifu Kaihatsu Enjo no Zaisei Mondai* [Financial Issues in Official Development Assistance]. In N. Z. Gakkai (Ed.), *Seifu Kaihatu enjo mondai no kento* [Examination of official development assistance]. (pp. 27–44).

Baishō Mondai Kenkyūkai [Reparations Research Group]. (1963). *Nippon no baisho* [Japan's reparations]. Sekai Janaru Sha (in Japanese).

DAC. (1962). *Annual Aid Review 1962: Memorundum for Japan*. OECD-DAC.

Fujimura, M. (1992). *The Untying of Japanese Aid: New Opportunities for Trade and Investment*. Economic Brief No.9. Private Investment and Trade Opportunities, East-West Center.

Hamada, K. (1950). *Ajia no Keizai saiken to nihon no yakuwari* [The reconstruction of economy in Asia and the role of Japan]. *Sekaikeizai, 5* (9), 19–27 (in Japanese).

Hays, S. (1971). *The Beginning of American Aid to Southeast Asia: The Griffin Mission of 1950*. Heath Lexington Books.

Hayashi, Y. (1959, November). *Nihon no Baisho Jisshino Genjyo to Jakkanno Kosatsu* [The State of Implementation of Japanese Reparation and Some Thoughts]. Ajia Kyokaishi, pp.10–18 (in Japanese).

Huq, M. (1995). *Reflections on human development*. Oxford University Press.

Imagawa, E., & Matsuo, D. (1973). *Nikka Haiseki: Ajia Minzokushugi to Nhihon* [Exclusionary movements against Japanese products: Asian nationalism and Japan). Nikkei Shinsho (in Japanese).

Imazu, T. (2009). *Nihon no gaiko seisaku to seifu kaihatsu enjo (ODA)* [Japan's foreign policy and Official Development Assistance (ODA)]. *Toa Keizai Kenkyu, 67*(2), 63–80 (in Japanese).

JICA (Japan International Cooperation Agency). (2019). *Kokusai Kyoryoku Kikoshi 1999–2018* [History of Japan International Cooperation Agency]. Kokusai Kyoryoku Kikou.

Hamana, H. (2017). *Jizoku Kanou na Kaihatsu Mokuhyo (SDGs) to Kaihatsu Shikin* [Sustainable development goals and development finance], Bunshindo (in Japanese).

Hunt, L. (2013, January 8). End of 30-year hunt for marcos billions?" *The Diplomat*. Asian Beat Section.

Katada, S. (2001). Why did Japan suspend foreign aid to China? Japan's foreign aid decision-making and sources of aid sanction. *Social Science Japan Journal, 4*(3), 39–58.

Keidanren (Nihon Keizai Dantai Rengokai). (2003). *ODA Taiko Minaoshi ni kansuru Iken* [Opinion on the Revision of ODA Charter]. Nihon Keizai Dantai Rengokai (in Japanese).

Kihara, M. (1966). *Sengo wagakuni no 'Keizai keikaku' no haikei: 'keizaikeikaku' to 'bouei seisan'ni tsuite* [On the background of "economic planning" in postwar Japan: With reference to "economic planning" and "defense production"]. *Keizai Ronso, 98*(3), 139–158 (in Japanese).

Kitaoka, S. (2015). Baisho mondai no seiji rikigaku [Political dynamics in the reparation issues]. *Kitaoka Shinichi, Monko Kaiho Seisaku to Nihon* (pp. 403–456). University of Tokyo Press (in Japanese).

Kindleberger, C. (1965). *Economic development* (2nd ed.). McGraw-Hill Book Co.

Kusano. A. (1989). *Taigaiseisaku kettei no kiko to katei* [Institutions and Processes of Foreign Policy]. Akio Watanabe, ed. *Kouza Kokusai Seiji 4 Nihon no Gaiko* (pp. 53–92). University of Tokyo Press.

Maemura, Y. (2019, November). *The historical development of Japanese ODA policy in the OECD: A corpus analysis of DAC aid reviews* (Japan's Development Cooperation: A Historical PerspectiveBackground Paper No. 9). JICA-RI. https://www.jica.go.jp/jica-ri/ja/publication/other/20180925_01.html

Lancaster, C. (2010). Japan's ODA-Naiatsu and Gaiatsu: Domestic sources and transnational influences. In D. Leheny & K. Warren (Eds.), *Japanese Aid and the construction of global development* (pp. 29–53). Routledge.

Mendoza, A. (2001). The industrial anatomy of corruption: Government procurement, bidding and award of contracts. *Public Policy, 5*(1), 43–71.

Ministry of Foreign Affairs. (1968). *OECD Kaihatsu Enjo iinkai 1968nen Tainichi Nenji Shinsakaigi Houkokoku* [OECD Development Assistance Committee 1968 Report of the Annual Peer Review on Japan]. Gaimusho.

Ministry of Foreign Affairs. (1973). *1973nen DAC Tainichi Nenji Shinsa* [1973 Annual Review of Japan by DAC]. Gaimusho (in Japanese).

Ministry of Foreign Affairs. (2007). *Ningen no Anzen hosho Panfuretto* [Panphet of Human Security]. Ministry of Foreign Affairs (in Japanese).

Ministry of Foreign Affairs. (2010). *Hirakareta Kokueki no Zoshin* [Promotion of national and worldwide interests]. Ministry of Foreign Affairs of Japan.

Ministry of Foreign Affairs. (2012). *Resolution of the general assembly of the United Nations on Human Security*. Available online https://www.mofa.go.jp/mofaj/press/release/24/9/0911_03.html. Accessed 19 September 2020.

Ministry of Foreign Affairs. (2020). *Nigen no Anzen hosho* [Human security]. Available online https://www.mofa.go.jp/mofaj/gaiko/oda/bunya/security/index.html. Accessed October 9, 2020.

Nobayashi, K., & Naya, M. (2015). *Kikigaki Ogata Sadako* [Oral history of Sadako Ogata]. Iwanami Shoten (in Japanese).

Okano, K. (1958). *Nihon Baisho Ron* [Theory of Japanese reparation]. Toyo Keizai (in Japanese).
Ōyama, T. (2018). Sengo nihon ni okeru 'kokueki" gainen no shingen: 'national interest' wo meguru honyakuronteki kousatsu [Origin of the Concept of "national interest" in Postwar Japan: Translational Analysis on the concept of "national interest"]. *Kokusai Anzen Hosho, 46*(3), 113–131.
Ōyama, T. (2019). Sengo nihon ni okeru ODA gensetsu no tenkankatei: rikoshugi tekinana kenchiwaikani zenkeika shitaka [The process of transformation in ODA discourse in Postwar Japan], *Nihon no Kaihatsu kyoryoku no rekishi background paper* No. 8. JICA Research Institute.
Pattajit, T. (2017). 1970 nendai no tai ni okeru hanichi undo: Sono genin no saikentou [Anti-Japanese movements in Thailand in the 1970s: Re-examining its causes]. *Soshio Saiensu, 25*, 45–61 (in Japanese).
Peason, L. (1969). *Partners in development: Report of the Commission on International development*. Pall Mall Press
Prime Minister's Secretariat. (1970). Eido antaiing ni tsuite [On aid untying]. *Zaisei Shouho*, No. 864, (pp. 12–15) (in Japanese).
Rivera, T. (2003). The politics of Japanese ODA to the Philippines, 1971–1999. In S. Ikehata, & L. N. Yu Jose (Eds.), *Philippines–Japan relations* (pp. 509–545). Ateneo De Manila University Press.
Shimomura, Y. et al. (1999). *ODA Taiko no Seiji Keizaigaku* [The Political Economy of ODA Charter]. Yuhikaku (in Japanese).
Sicat, G. (2014). *Cesar Virata: Life and times of four decades of economic history in the Philippines*. University of the Philippines Press.
Sumi, K. (1989). *ODA Enjo no genjitsu* [Realities of ODA]. Iwanami Shinsho (in Japanese).
Takehara, N. (2014). *Nihongata ODA to Zaisei* [Japanese Style ODA and Fiance]. Mineruva Shobo (in Japanese).
Tanaka, Y. (2003). Seifu Kaihatu Enjo Taiko no Kaitei wo megutte [On the amendment of official development assistance charter]. *Osaka Jogakuin Tankidaigaku Kiyo, 33*, 113–129 (in Japanese).
Ueda, S. (2010). 'Ningen no Anzen hosho' no Hatten [Development of "human security"]. *Sandai Hogaku, 44*(2), 1–22 (in Japanese).
Udagawa, M. (2017). "Nihon no enjo rinen toshiteno jijodoryokushien no kokusai seijirironteki keifu: Ningenno anzenhosho tono kankeisei wo shiyani (Assistance for Self-help Efforts as Japanese Aid Principle in the context of International Political Theory: Focusing on the Connection between Human Security). *Kokusai Seiji, 186*, 113–128 (in Japanese).
Yanagihara, T. (2019). 'Ningen no Anzen Hosho' ni miru nihon no enjo no Tokushoku: Gaimusho, JICA bunsho no Rebuy yori (Characterizing Japanese Aid through "Human Security": Reviewing documents from Ministry of

Foreign Affairs and JICA) *Nihon no Kaihatsu kyoryoku no rekishi background paper* No .6. JICA Research Institute (in Japanese).

CHAPTER 4

South Korea's Foreign Aid as a Foreign Policy Instrument

Jisun Song and Eun Mee Kim

1 Introduction

One of South Korea's most notable characteristics is its successful transition from a recipient country that once received aid to a donor country that provides aid to other developing countries. Starting from the liberation in 1945, South Korea received official development assistance (hereafter, ODA) from major bilateral donors such as the United States (hereafter, the US) and Japan as well as from multilateral institutions including the United Nations (hereafter, the UN). It was only in 1995 and 2000 that South Korea officially graduated from the recipient list of

J. Song (✉)
Korea National Diplomatic Academy, Seoul, Republic of Korea
e-mail: jisunsong01@gmail.com

E. M. Kim
Ewha Womas University, Seoul, Republic of Korea
e-mail: emkim@ewha.ac.kr

H.-j. Kwon et al. (eds.), *International Development Cooperation of Japan and South Korea*,
https://doi.org/10.1007/978-981-16-4601-0_4

the World Bank and the Organisation for Economic Co-operation and Development (hereafter, OECD), respectively.

Against this backdrop, this chapter will analyze South Korea's international development cooperation as a donor, with special reference to the history of policies and institutional structure. We will analyze the major changes in policy and strategic objectives of South Korea's ODA while taking into account key factors that influenced major policy changes. Second, the change in government institutional structure in terms of ODA management and implementation will be reviewed. Third, the history of South Korea's ODA disbursement will be examined in detail (e.g., by sector, aid type, recipients, etc.) to analyze the effect of the policy changes. The timeframe of the analysis will be from 1963 to March 2021.

2 Foreign Aid as a Foreign Policy Tool

The debate on whether foreign aid is an instrument of foreign policy or an end in itself has continued since its inception. The flow of resources from one country to another innately forms a relationship between the two countries, sometimes even leading to dependency. While foreign aid is in principle a tool used by developed countries to support developing countries, the underlying functions have often been more complex, serving the donor country's national interest whether political or economic. For instance, while aid was a diplomatic expedient during the Cold War era, it has shifted to a tool that centers around developmental motivations since 2000. Also, aid from former colonizers to former colonies reinforces the diplomatic and political alliance between those countries (Apodaca, 2017). Not only do multiple motivations coexist, but foreign aid is also influenced by domestic institutional factors such as bureaucratic incentives at times (Arel-Bundock et al., 2015).

Within this context, developed countries have utilized foreign aid as a diplomatic tool for decades. Foreign aid can function to maintain or even change the status quo between countries by being a component of donors' foreign policy portfolio. The more powerful the donor is, the more likely that it will dedicate more resources to seek changes than weaker countries (Palmer et al., 2002). Some scholars argue that donors' national interests determine the provision and characteristics of their foreign aid, including the timeframe, channel, sector, type (such as bilateral or multilateral), and form (such as grants or loans) (Apodaca, 2017). Morgenthau (1962) goes as far as to argue that all forms of foreign aid are political with the exception of humanitarian aid, although it is not completely free from politicization when provided in a political context. Empirically, major donors such as the US have utilized foreign aid as an "essential instrument of US foreign policy" that has become an integral part of the US's national security strategy since 2000 (Tarnoff & Lawson, 2016), and effective foreign aid programs "can and should be a crucial component" of diplomatic strategy (Atwood et al., 2008).

The authors also posit that South Korea's foreign aid has been implemented as part of its foreign policy. The authors will first review the foreign policies of each decade, link them to the foreign aid policies of the corresponding period, and analyze whether and how foreign aid was implemented in line with diplomatic strategies.

3 Analysis

3.1 1960s: Beginning of South Korea's Aid Amid the Cold War

South Korea's foreign affairs strategy of the early 1960s was shaped by changing international relations primarily driven by the emergence of neutral countries in Asia and Africa as well as the shifting relationship between the United States (US) and the Soviet Union. In a way, not only was South Korea too weak to devise an independent foreign strategy, but South Korea's close relationship with the West also left little choice to suggest otherwise (Koo, 1980). As a result, the South Korean government adopted active diplomacy as the basic policy of foreign affairs and sought to diversify its foreign relations based on elevated national power. For instance, South Korea formed diplomatic relations with non-aligned, neutral countries including Burma (now Myanmar), Indonesia, Nepal, and Pakistan, while dispatching goodwill envoys in large numbers to such countries. Also, South Korea aimed to deepen diplomatic ties by going beyond a focus on goodwill to cover areas of the economy, culture, and more (MOFA, 1971, 2009). President Park Chung Hee's (1963–1979) State of the Union message to the National Parliament in 1965 reflects the fundamental direction of South Korea's foreign policy and foreign aid of this era (Park, 1965).

> We cannot overlook the fact that Communist China and its followers are posing an increasing threat to Southeast Asia, so we will always keep a close eye on the region... Also, from 1965 onwards, we will concentrate our efforts in Africa to ensure and enhance our country's status in the region. Specifically, we will increase diplomatic missions and at

> times dispatch goodwill envoys, thereby carrying out diplomacy based on the African continent. Furthermore, we will actively strengthen technical assistance projects.... (authors' translation)

Subsequently, the overall direction of pursuing active diplomacy, particularly in terms of engaging with the member countries of the non-aligned movement, was also reflected in foreign aid. All the countries to which South Korea provided assistance in the 1960s were non-aligned countries (see Table 1). While South Korea also practiced aid as a member of the Colombo Plan,[1] the objectives of foreign aid were vague. Foreign strategy and therefore foreign aid were heavily influenced by the need for national security during the Cold War era.

Since South Korea did not have a big foreign aid budget, aid was largely in the form of small grants and technical cooperation. Only a select number of ODA projects were implemented in the 1960s, and there was no specific ministry or agency in charge of aid. Although it is unclear whether South Korea's aid projects were in line with the OECD DAC's definition of ODA,[2] they were categorized as *wonjo* (aid) projects. Bilateral aid was mainly administered by the Ministry of Foreign Affairs (MOFA), while the dispatch of experts and invitations for training were also handled by the Ministry of Science and Technology (MOST). MOST was also responsible for some of the technical assistance, particularly training programs held in South Korea. While the aid projects in the first half of the 1960s were mainly in the form of invitations for training, South Korea would begin to provide medical technical assistance to partner countries in Africa from 1968 onwards. As of December 1970, South Korea had dispatched 28 doctors to 13 partner countries in Africa in the form of technical assistance combined with the provision of medical supplies (MOFA, 1971).

Although South Korea initiated its international development cooperation with training programs, a lack of financial resources prevented the

[1] The Colombo Plan for Cooperative Economic and Social Development in Asia and the Pacific was officially launched in 1951 and is a regional intergovernmental organization for the furtherance of the economic and social development of the region's nations. South Korea became a member in 1962 (Colombo Plan, 2020).

[2] OECD DAC defines ODA as government aid that promotes and specifically targets the economic development and welfare of developing countries (*Source* OECD DAC website).

Table 1 South Korea's bilateral aid in the 1960s

Year	*Country*	*Content*	*Ministry*
1962	Laos	• Provide humanitarian assistance in the form of medical supplies, school supplies, and clothing via the Red Cross upon request for aid for refugees	MOFA
1967	Lesotho	• Provide 2 Willys Jeeps (automobile)	MOFA
1968	Benin (Dahomey)	• Agree to dispatch 1 physician and 1 surgeon, 1 incubator, 1 ambulance, and antibiotic and anti-tuberculosis drugs	MOFA
	Côte d'Ivoire	• Agree to dispatch 1 physician and 1 surgeon	MOFA
	Ethiopia	• Agree to dispatch 1 physician and 1 surgeon	MOFA
	Gambia	• Agree to dispatch 3 physicians and surgeons	MOFA
	Niger	• Agree to dispatch 3 physicians and surgeons, ENT equipment, 2 ambulances, and antibiotic and anti-tuberculosis drugs	MOFA
	Rwanda	• Dispatch medical experts	MOFA
1969	Democratic Republic of Congo	• Dispatch 2 pediatricians	MOFA
	Lesotho	• Provide medical supplies of USD 10,000	MOFA
	Gabon	• Agree to dispatch 1 physician and 1 surgeon	MOFA
	Rwanda	• Agree to provide 4 ambulances (around USD 20,000)	MOFA
	Sierra Leone	• Agree to dispatch 1 ophthalmologist and 1 radiologist	MOFA
	Togo	• Agree to dispatch 1 physician and 1 surgeon	MOFA
	Vietnam	• Provide technical assistance of USD 300,000 for development of the Mekong region	MOFA

Note At times, the actual dispatch of experts took place one or two years after the agreement was made. For instance, although South Korea agreed to dispatch one physician and one surgeon to Côte d'Ivoire in 1968, they were actually dispatched in 1969
Source J. H. Kim (2014), KOICA (1997), and MOFA (various years)

Table 2 Number of invited trainees by funding source

Year	*Total*	*South Korea*	*With USAID*	*w/ UN or others*
1963	8	–	8	–
1964	19	–	19	–
1965	46	6 (USD 10,993)	40	–
1966	34	4 (USD 6,435.50)	30	–
1967	90	3 (USD 7,857.80)	83	4
1968	122	6 (USD 11,051)	114	2
1969	134	15 (USD 49,144.90)	109	10
1970	177	15 (USD 12,187.50)	158	4

Source J. H. Kim (2014) and MOFA (1971)

government from implementing them independently. As a means to overcome the budget constraint, South Korea first implemented its foreign aid project in 1963 with a financial contribution from the USAID. It was only from 1965 that the South Korean government implemented training programs that were fully funded by the government. This shows that the budget input from the South Korean government had gradually increased from around USD 10,000 in 1965 to almost USD 50,000 in 1969. Whereas the training programs jointly funded with the USAID and UN invited participants from Asia (J. H. Kim, 2014), those funded by the South Korean government invited trainees from both Asia and Africa, such as from Nepal, Liberia, Somalia, Swaziland, Thailand, and Uganda (MOFA, 1971), all of which are members of the non-aligned movement. This indicates a close connection between foreign policy and foreign aid (Table 2).

3.2 1970s: Diplomatic Competition with North Korea as the Driver of South Korea's Aid

South Korea's diplomacy in the 1960s and 1970s was guided by the government's priority placed on peace and national security, as well as claiming legitimacy at home and abroad (Koo, 1980; Lee, 1978). The trend also became more evident in the 1970s when many countries in the Global South became UN member states and non-aligned states transformed into a formidable political force. North Korea had already begun offering foreign assistance (e.g. grants, technical cooperation) in the 1960s and obtained observer status in the UN in 1973 (Chae, 2014). North Korea had also achieved an economic growth rate of 13.7% in 1956–1960 and 4.1% in the 1960s (Cho & Kim, 2020, p. 32), fueling the competition. While these numbers may have posed threats to South Korea, the 1970s was also the time when South Korea was undergoing rapid economic growth, reaching an average GDP growth rate of 10.5% (Statistics Korea, 2021).

Such a context also affected South Korea's foreign policy in the 1970s, which centered around the three ideas of security diplomacy, diplomacy to foster a basis for unification, and economic and cultural diplomacy. The government maintained active involvement with neutral countries as it had done in the previous decade, dispatching special envoys every year to countries in various regions to stress international support and mutual cooperation in terms of South Korea's unification diplomacy policy and peace on the Korean peninsula, in order to have an upper hand in diplomacy over North Korea (MOFA, 2009).

There was also economic motivation behind South Korea's diplomacy, albeit to a lesser extent. According to the report submitted by the Korea National Diplomatic Academy in 1972, there was a need to build a better awareness of diplomacy with countries in Africa by shifting away from UN-centered diplomacy to improve favorable relations with countries in Africa. The report suggested that South Korea's diplomacy go beyond responding to North Korea's international activities to also consider improving South Korea's trade, economic advancement, and resource development. It also called for expanding economic diplomacy, invitation and visitation, and public relations activities. To this, the Ministry of Foreign Affairs of South Korea responded that in the 1970s the government secured a comparative advantage in competition with North Korea in the short run while further solidifying economic diplomacy with countries in Africa so that it could contribute to South Korea's economic development in the long run. It also suggested refraining from unnecessary aid competition with North Korea (MOFA, 1971).

Foreign policy continued to guide South Korea's foreign aid in the 1970s as it did in the 1960s. The two Koreas sought to strengthen their respective allies, particularly in Africa, to win more votes in the UN, and foreign aid was utilized as a diplomatic means to reach that goal. This was reflected in President Park Chung Hee's administrative policy speech at the National Assembly in 1976, in which he remarked that "we will exert multifaceted efforts for all other countries to deeply understand our peace ideas and unification policy and to support our position in the international community" (Park, 1976).

Table 3 shows South Korea's ODA disbursed between 1970 and 1975. Compared to foreign aid provided in the 1960s, the number of South Korea's partner countries and the scope of the foreign aid budget and materials both rose. While South Korea's aid in the previous decade was mainly in the form of training and medical technical cooperation, South Korea began to provide humanitarian aid in the form of cash, dispatching experts to developing countries, and commodity aid. Between 1973 and 1975, South Korea provided around USD 1,705,500, which consisted of USD 1,118,600 in financial aid to 16 countries and USD 586,900 in commodity aid to 13 countries in Africa (MOFA, 1975).

South Korea's strategic use of foreign aid to fulfill its diplomatic objectives was sustained until the late 1970s. For instance, the number of countries that the MOFA provided grant aid to gradually increased from 15 countries in 1978 to 25 countries in 1979 and 28 countries in 1980 (OPM, 1979, 1980). Technical cooperation such as dispatching experts and invitations for training, as well as the provision of commodities and emergency relief to Latin American countries, also began during the 1970s (J. H. Kim, 2014), including Jamaica, Grenada, and Panama as mentioned in the Table 3.

Despite such international development cooperation efforts, 30th UN General Assembly held in 1975 adopted two resolutions, with one each endorsed by South Korea and North Korea. The resolution that sided with South Korea was adopted with 59 votes to 51, and that which sided with North Korea was adopted with 54 votes to 43 (Choi, 1984). This voting turnout further sparked South Korea to better utilize foreign aid to gain more support in global platforms. Even after the UN General Assembly, North Korea continued to provide more grants to more countries than South Korea during this period (see Table 4). For instance, South Korea provided less than half of what North Korea provided between 1976 and 1978. It was only in 1979 that South Korea reached out to more countries than North Korea while narrowing the foreign aid budget gap.

Table 3 South Korea's ODA in the early 1970s

Year	*Country*	*Content*
1970	Burkina Faso	• Dispatch 2 surgeons and physicians
	Rwanda	• 4 ambulances (worth USD 2 million)
	Lebanon	• Aid of USD 3,000 for refugee relief and diplomatic ties
	Jordan	• 1 guitar and 2 radios
	Swaziland	• Send 2 physician and surgeons
1971	Palestine	• Relief aid of USD 5,000 via UNRWA
	Lebanon	• Relief aid of GBP 10,000 (USD 3,000 and GBP 520)
	Cambodia	• 10,000 military uniforms (worth KRW10,080,000)
1972	Niger	• Product for construction
	Lebanon	• Relief aid of USD 5,000 and 1 ambulance
	Vietnam	• 120,960 ramen (instant noodles) for emergency relief (worth USD 5,000)
	Iran	• Medical supplies for emergency relief (worth USD 5,000)
	Senegal	• Medical supplies, 2 ambulances
	Togo	• 1 x-ray, 100 hospital beds
	Gambia	• Dispatch of 2 doctors (as medical experts)
	Cambodia	• Medical supplies
1973	Chad	• Relief aid of USD 10,000 and 5 automobiles (worth USD 33,000)
	Nicaragua	• Aid of USD 20,000
	Central African Republic	• Dispatch of 2 agriculture experts and 1 Taekwondo expert
		• 14 ambulances (worth USD 70,526)
	Niger	• Relief aid of USD 10,000
	Lebanon	• Relief aid of USD 7,000
	Tunisia	• Relief aid of USD 10,000
1974	Nicaragua	• Sport shirts worth USD 5,000
	Egypt	• Relief goods (including medical supplies) (worth USD 39,000)
	Central African Republic	• Aid of USD 1,370 to support construction of competition area
	Malawi	• USD 20,000 as part of medical cooperation
	Lesotho	• Economic cooperation fund of USD 50,000
	Myanmar	• Relief aid of USD 1,000
1975	Rwanda	• USD 100,000 to build 10 school buildings

(continued)

Table 3 (continued)

Year	*Country*	*Content*
	Laos	• Invitation of 10 trainees • 50 yards of textile materials
	Cambodia	• Medical supplies (worth USD 4,372,000)
	Panama	• 30 Cultivators
	Lesotho	• 1 printer (worth USD 12,000), 2 tractors (worth USD 18,000) • Medical cooperation fund of USD 20,000
	Burkina Faso	• 4,000 military uniforms
	Mauritius	• Relief aid of USD 10,000
	Grenada	• 2 ambulances and 20 cultivators (worth USD 52,255) • Medical supplies (worth USD 10,000)
	Jamaica	• 20 cultivators (worth USD 39,532)
	Central African Republic	• USD 3,574 to hold bazaar and USD 10,000 for construction • Dispatch 1 design expert
	Chad	• Military uniforms and 2,000 pairs of military shoes
	Turkey	• Relief aid of USD 10,000 and medical supplies (worth USD 10,000)
	Fiji	• Relief aid of USD 10,000

Source MOFA (Various years)

Table 4 Grant from South and North Korea (unit: USD 10,000)

	1976	*1977*	*1978*	*1979*
South Korea	90 (11)	110 (14)	135 (19)	182 (24)
North Korea	317 (22)	435 (22)	510 (25)	350 (15)

Note The number in parenthesis refers to the number of recipient countries
Source Kim (2014)

3.3 1980s: Increasing Demand for South Korea's Contribution to Global Development

The 1980s marked the shift in South Korea's diplomatic principles to a stronger concentration on geopolitical and economic gains in the global arena. On the one hand, South Korea's foreign aid was re-driven by security concerns and competition with North Korea. North Korea strengthening its diplomatic activities in this era led South Korea to seek closer and stronger security cooperation with its allies (MOFA, 2009). On the other hand, rapid economic growth led South Korea to be deeply integrated into the global market, including with Southeast Asian countries. Not only did South Korean private firms increase direct investment in the region for abundant and cheap labor, but it was also a niche market for South Korean companies that were undergoing trade conflict with developed countries. Priority focus on Asia would continue from the 1980s onward for such commercial reasons (Chun, 1996).

Such political and economic purposes were well-reflected in South Korea's foreign policy and thereby foreign aid. Upon taking office in March 1981, President Chun Doo Hwan (1980–1988) mentioned in his administrative policy speech to the National Parliament in October of the same year that the basis of foreign policy, including security, the economy, and resources, as well as peace and unification diplomacy, would be based on autonomy, activeness, and interests. He also stressed that the government would strive to improve cooperative relationships across various sectors with its allies in Asia, the Pacific, Africa, the Middle East, and Latin America. In particular, the relationship with the Association of Southeast Asian Nations (ASEAN) member states as well as non-aligned countries would be augmented while continuously applying the open-door policy to countries whose ideology and systems differed from South Korea. Economic and technical cooperation with foreign countries would be further expanded (Chun, 1981). The Chun administration sought to deepen relations with developing countries by inviting high-level officials and visiting relevant countries, consolidating South Korea's comparative advantage in diplomacy over North Korea, and fostering conditions for friendly relations with countries that South Korea had not had diplomatic ties with (GOV, 1986). It also aimed to apply south–south cooperation with developing countries by supporting South Korean private companies' participation in developing countries' economic development projects, promoting a complementary and cooperative relationship between developing countries' resources and South Korea's technology, and diversifying diplomatic efforts such as boosting cultural and sports exchange (GOV, 1986).

The 1980s was also the time when South Korea displayed a stronger presence as an aid donor. The 1986 Asian Games and the 1988 Seoul Olympics contributed to reducing foreign debt and creating a surplus balance in international trade, which motivated the government to fully transfer from an aid recipient to a donor. There was also a growing global request for South Korea to allocate more global aid owing to its positive economic development and the domestic need to disburse more foreign aid to better venture into the global markets (KOICA, 1997).

In 1983, MOFA reviewed the outcome of South Korea's grant aid in the previous year and determined that grant aid was effective in accomplishing diplomatic goals in non-aligned countries, deterring North Korea from forming sole diplomatic relations, laying the groundwork for relations with countries with no diplomatic ties, and securing support for South Korea at the 31st UN General Assembly. Based on this review, the fundamental objectives of South Korea's grant aid strategy in 1983 were to form diplomatic ties with non-aligned countries and to gain support in the UN, sustain North Korea's maneuvers, advance relations with countries with no diplomatic ties, implement south–south cooperation, and promote economic and commercial cooperation. These goals would be accomplished by donating items produced in South Korea, providing emergency relief and special support for refugees, inviting trainees, and dispatching experts overseas. All of this would be achieved with an annual grant budget of USD 7 million (MOFA, 1983).

Table 5 shows the composition of South Korea's ODA between 1984 and 1989. South Korea's total ODA in 1984 was around USD 28 million but would more than double to reach USD 63.8 million in 1989. Interestingly, multilateral ODA accounted for a larger share of South Korea's ODA during this time. For instance, multilateral aid made up around 56% of total aid in 1984, and this number would rise to 64% in 1989. This is contrary to the current style of South Korea's aid whereby multilateral ODA is generally estimated to be around 25% of total aid (CIDC, 2020a). Multilateral ODA was mainly composed of annual contributions to international organizations (H. S. Kim, 1990). South Korea mainly practiced grant aid and technical assistance during this period, and concessional loans only began from 1989. Even in 1989, the concessional loan portion was less than one-tenth of the grant aid, which is strikingly different from concessional loans accounting for 46% of total aid in 2019 (CIDC, 2020a).

The 1980s was also when more ministries and government agencies got involved in implementing ODA. For instance, it can be inferred from Table 6 that whereas MOFA and MOST were the two main ministries in charge of grant aid and technical assistance, the Ministry of Labor (MOL)

Table 5 South Korea's ODA in the 1980s (unit: USD million)

	1984	*1985*	*1986*	*1987*	*1988*	*1989*
Total ODA	28.0	48.3	109.5	73.0	55.0	63.8
Grant aid	6.3	8.5	9.1	11.4	12.6	12.2
Technical assistance	6.1	6.4	6.5	8.2	8.5	8.5
- Training	2.3	2.4	2.2	2.3	2.9	3.6
- Expert	0.2	0.3	0.3	0.6	0.7	0.7
- Consultation	1.3	1.2	1.3	0.9	1.1	1.9
- Project	-	-	-	1.4	0.9	0
- Taekwondo, medical experts	1.7	2.1	2.1	2.3	2.4	1.7
- Others	0.6	0.4	0.6	0.7	0.5	0.6
Loans	–	–	–	–	–	1.8
Multilateral ODA	15.6	33.4	94.5	55.7	35.1	40.9
ODA/GNP (%)	0.03	0.05	0.11	0.06	0.03	0.03
Other official flows (OOF)	359.9	368.0	275.6	346.5	433.0	304.8
ODA + OOF/GNP (%)	0.45	0.46	0.37	0.33	0.29	0.18

Source Kim (1990)

would take over the training program from MOST and implement it from 1983 onwards. Other ministries such as the Ministry of Construction would also participate in training programs for the construction sector from 1983 onwards and implement actual construction projects. The Ministry of Communications would also initiate grant projects in the corresponding sector. The Korea Advanced Institute of Science and Technology (KAIST), a leading South Korean university for science and technology, implemented co-research projects in 1981. Also, the Korea Development Institute (KDI) kicked off the International Development Exchange Program (IDEP) from 1982 onwards. IDEP is a technical cooperation program with the goal of increasing understanding of important development issues, fostering the exchange of ideas on certain policy issues, and strengthening partnerships for better understanding and cooperation with representatives from developing countries (KDI, 1983).

In 1987, the Economic Development Cooperation Fund (hereafter, the EDCF) in charge of concessional loans was formally established based on the Foreign Economic Cooperation Fund Act enacted in December 1986. Article 7 of this act states that the fund will be used to promote the industrial development of developing countries as well as an economic interchange between South Korea and the partner countries to stabilize partner countries' national economies, among other goals (NLIC,

Table 6 Key milestones of South Korea's ODA and managing institutions

Timeline	*Year*	*Content*	*Managing institution*
1960s	1963	• Invite people from developing countries for training for the first time with financial contribution from USAID	MOFA, MOST
	1965	• Implement invitation for training with government resources	MOFA, MOST
	1967	• Dispatch experts with government resources • Implement invitation for training in partnership with UN and other international organizations	MOFA, MOST
	1969	• Implement technical assistance projects with government resources	MOST
1970s	1975	• Start invitation for training program for technical engineers – MOST to implement the training program from 1975 to 1982 – MOL to be in charge from 1983 onwards	MOL
	1977	• Start commodity aid with equipment	MOFA
1980s	1981	• Start co-research projects	KAIST
	1982	• Start International Development Exchange Program (IDEP)	KDI
	1983	• Start invitation for training program for construction technician experts	MOC
	1984	• Start construction technology projects (grant) • Start a project that supports establishment of vocational training facility	MOC, MOL

(continued)

Table 6 (continued)

Timeline	*Year*	*Content*	*Managing institution*
	1987	• Establish the Economic Development Cooperation Fund (EDCF) • National Research Foundation (NRF) to start implementing overseas technical cooperation	MOF, MOST
	1988	• Start communications and technology projects for developing countries (grant)	MOCOM
	1989	• Start dispatching Korean volunteers overseas • Start technology projects (grant)	Korean National Commission for UNESCO, EPB
1990	1991	• Establish the Korea International Cooperation Agency (KOICA)	Under MOFA
	1995	• Start dispatching international cooperation agents • Start support projects for non-governmental organizations	KOICA

Note MOC stands for the Ministry of Construction, MOL for the Ministry of Labor, MOCOM for the Ministry of Communication, MOF for the Ministry of Finance, EPB for the Economic Planning Board, KDI for Korea Development Institute, and KAIST for the Korea Advanced Institute of Science and Technology
Source KOICA (1997)

1986). While the EDCF was placed under the Korea Export–Import Bank (Eximbank), the fund management council was established within the Economic Planning Board (EPB) with the head of EPB serving as the council chair. The council would deliberate on the policy direction as well as the programs of the EDCF (MOFA, 1987). It also handled the selection of the EDCF's priority partner countries (MOFA, 1989). EPB was later integrated within the Ministry of Finance in 1994 and became the Ministry of Economy and Finance (MOEF) (then, the Ministry of Finance and Economy). Thus, it can be inferred that the EDCF was heavily influenced by the decision makers of the EPB and subsequently by the finance ministry and that the EDCF was implemented in line with the government's economic agenda. Within MOFA, the Economic Cooperation Division under the Economic Affairs Bureau was mainly in charge of international development cooperation (MOGA, 1986), similar to the roles that the three divisions of the Development Policy Division, the Development Cooperation Division, and the Multilateral Development Cooperation/Humanitarian Assistance Division under the Development Cooperation Bureau play today.

3.4 1990s: Institutionalization of South Korea's ODA

The end of the Cold War signaled shifts in the global setting for South Korea, with the rise of global protectionism and the foundation of the World Trade Organization, withering competition with North Korea, and growing demand for South Korea's role as a member of the international community. This implied prioritization of economic expansion over national security, diversification of its diplomatic partners (especially in Asia), and enhancing contributions to other developing countries (Chun, 1996).

Accordingly, South Korea began to highlight its ability to act as a bridge between developed and developing countries as a middle-power country in global platforms. For instance, President Roh Tae Woo (1988–1993) mentioned in his keynote speech at the UN General Assembly in 1991 that "South Korea, as a middle country between developed and developing countries, will play an active role to solve problems between Global South and North, and thereby repay what it received." Specifically, "South Korea will not only share its experience and technology but also act as a bridge to accelerate the exchange of and cooperation in the capital, market, and information" (Roh, 1991). The subsequent President Kim Young Sam (1993–1998) also remarked that South Korea was endowed with the role to "coordinate and harmonize countries in different stages of economic development" (Y. S. Kim, 1994), will "increase its ODA to the point that is equivalent to its economic capacity" (Kim, 1995a), and that South Korea emerged as a "middle-power country within the UN only four years after becoming a member" (Kim, 1995b). Within this context, there was an understanding that South Korea should convey its own development experience and know-how to support developing countries' socioeconomic development and implement a corresponding style of foreign aid (KOICA, 1997).

In line with the commitment, South Korea strived to gradually increase its ODA budget while institutionalizing governmental agencies for grants and loans. It aimed to raise ODA to 0.1% of the GNP by 2000 (KOICA,

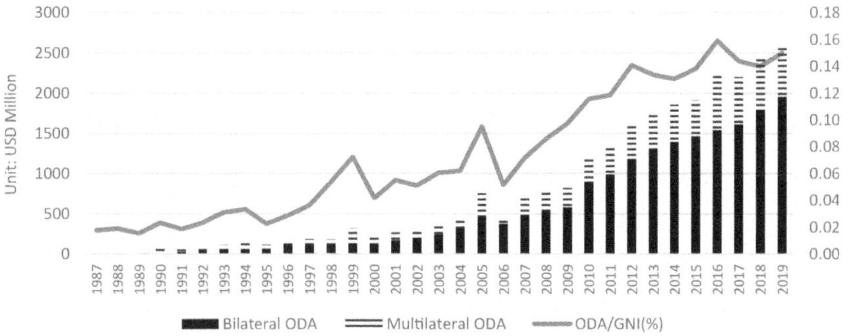

Fig. 1 South Korea's ODA (1987–2019) (unit: USD million) (*Note* Disbursement and current price based. *Source* OECD International Development Statistics [accessed 2 November 2020])

1997), although it only ended up reaching 0.04% in 2000. Regardless, South Korea's total ODA grew from USD 23.5 million in 1987 to USD 61.2 million in 1990 and USD 317.5 million in 1999. Figure 1 shows that not only did South Korea's bilateral, multilateral, and total aid grow over time, but even the ODA/GNI ratio (%) rose from 0.02% in 1987 and 1990 to 0.07% in 1999.[3]

South Korea's aid became more concentrated in Asia than Africa starting in the 1990s (see Fig. 2). For instance, whereas South Korea provided USD 5.3 million in bilateral assistance to Africa and USD 13.6 million to Asia in 1991, the gap expanded further, with Africa receiving USD 10.2 million and USD 106.5 million in 1999. As was the case in the 1980s, Asia was a strategic partner for South Korea in both economic and political terms, with geographic proximity, abundant natural resources, a growing market, and growing investment in the region by the South Korean private sector. It was also crucial for South Korea to build a stronger alliance with the region to receive support from international organizations (Chun, 1996). Prioritization of Asia is a trend that would continue in the 2000s, with President Kim Dae Jung (1998–2003) stating in 2001 that the South Korean government would consistently enlarge

[3] Yet, this was far below the DAC average of 0.33% ODA/GNI in 1987, 0.32% in 1990, and 0.22% in 1999.

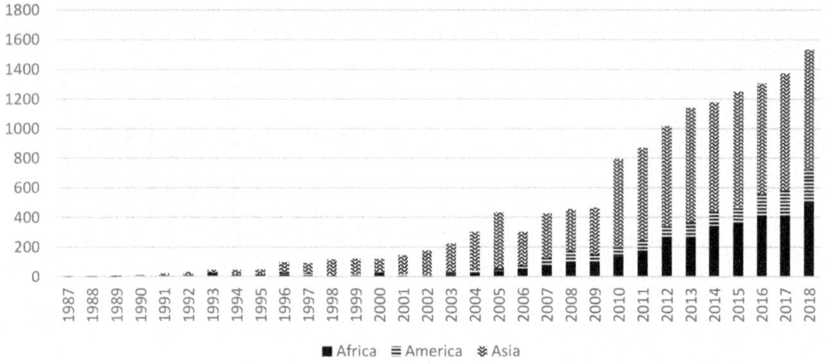

Fig. 2 South Korea's ODA by region (1987–2018) (unit: USD million) (*Source* OECD International Development Statistics [accessed 2 November 2020])

the share of grants and loans to Asia, particularly the ASEAN region (Kim, 2001).

South Korea's ODA system became further sophisticated in the 1990s with the establishment of the Korea International Cooperation Agency (hereafter, KOICA) in 1991 as the main grant aid agency under the supervision of MOFA. Then, there was rising concern related to the fragmented aid system that ultimately led to aid ineffectiveness. KOICA was born out of such a concerted effort to have one agency that was wholly in charge of the government's international development cooperation (KOICA, 1997).

When established, KOICA was to be in charge of integrating and managing diverse technical assistance and human resource exchange projects that were sporadically implemented by MOST, MOC, and other ministries (KOICA, 1997). MOFA would be in charge of grants and contributions to UN agencies, whereas MOEF would supervise all concessional loans and contributions to international financial institutions such as the World Bank. This system has remained the same in the 2000s, although now over 40 ministries and agencies are participating in implementing grant aid and providing contributions to certain international organizations relevant to their work area (Table 7).

Table 7 South Korea's ODA management system in the 1990s and 2000s

Financial resource	*ODA type*			*Implementing agency*	*Supervising ministry*
ODA	Bilateral ODA	Grant	Project, technical cooperation, food aid, emergency relief, etc	KOICA	MOFA
		Loan	Concessional loans	EDCF	MOEF
	Multilateral ODA	Contributions to international financial institutions			MOEF
		Contributions to the UN and other agencies			MOFA MOEF
OOF	Foreign investment				

Source KOICA (1997)

3.5 2000s: Solidification of South Korea's Position as a Bridge Between Global North and South

South Korea's aspirations to solidify its bridging role between the Global North and South while promoting economic gains is very valid and reflected in its foreign policy. However, a distinctive feature of South Korea's ODA in the 2000s is that although guided by foreign policy, the government would eventually establish specific priority goals and strategies for ODA alone. The Framework Act on International Development Cooperation (Act No. 9938), enacted in 2010, lays out the basic principles of South Korea's ODA to reduce poverty, improve quality of life, promote development and systematic changes, and pursue friendly ties and mutual exchanges with developing countries. Although each administration devised different strategies for ODA as explained below, all worked to achieve the 0.20% ODA/GNI by 2020 and 0.30% ODA/GNI by 2030.[4] To support its stronger commitment, South Korea went through key institutional improvements.

In terms of foreign strategy, President Roh Moo Hyun (2003–2008) produced the Improvement Plan in Korea's ODA Policies to Developing Countries in 2003, the Comprehensive ODA Improvement Plan in 2005, Korea's Initiative for Africa's Development in 2006, and the Mid-Term Strategy for Development Cooperation in 2007. In particular, the Initiative for African Development announced the intention to triple South Korea's aid to Africa by 2008 (MOFAT, 2006). While this did not result in the reversal of priority regions, it reflected a powerful commitment to prioritize poverty and development as well as to expand relations with Africa through foreign aid. President Lee Myung-bak (2008–2013) announced Global Korea as the pillar of the national development strategy. This Global Korea strategy contains the following five tasks: develop a new peace system, practice pragmatic diplomacy including resource diplomacy, advance security, promote green growth, and enhance cultural power (GoK, 2008). In particular, the administration included the goal to reach 0.15% ODA/GNI by 2012, increase multilateral and humanitarian assistance, advance the ODA implementation system, and strengthen diplomacy with Africa as part of the national

[4] However, South Korea failed to reach the 0.20% ODA/GNI by 2020, reaching only 0.14% (OECD, 2021).

policy task to contribute to solving global issues (GoK, 2011). The subsequent President Park Geun-hye administration (2013–2017) announced four development cooperation initiatives that focused on specific sectors and stakeholders: Better Life for Girls, Safe Life for All, Science, Technology, and Innovation for a Better Life, and the New Rural Development Paradigm.[5] At the same time, the overarching goal was to reach 0.20% ODA/GNI by 2020. The incumbent President Moon Jae-in (2017–2022) announced the New Southern Policy and the New Northern Policy, two major foreign policies focusing on southern and northern Asia, respectively. It is within these foreign policies that the government practices ODA.

The 2000s was when South Korea advanced the institutionalization of ODA. It formed the Committee for International Development Cooperation (CIDC) in 2006 to tackle the aid fragmentation problem and applied to become an OECD DAC member in 2007 (GOV, 2017). In 2010, South Korea enacted the Framework Act on International Development Cooperation to set the legal basis of South Korea's ODA while devising the Strategic Plan for International Development Cooperation in the same year. It also formulated the five-year Mid-term Strategy for International Development Cooperation (hereafter, the mid-term strategy) for 2011–2015 and 2016–2020. Since 2011, CIDC has also approved annual international development cooperation action plans that lay out the goals and policy directions of South Korea's ODA each year while also producing the first Country Partnership Strategy (CPS) for priority partner countries for 2011–2013 and subsequently the second CPS for 2016–2020. South Korea also hosted the G20 Seoul Summit in 2010 and the Fourth High-Level Forum on Aid Effectiveness in Busan in 2011. This indicates that South Korea has actively implemented a series of policies and strategies to enhance aid coordination, coherence, and effectiveness. It also reflects the recommendations provided by the OECD DAC's special review in 2008 as well as peer reviews in 2012 and 2018 (GOV, 2017).

South Korea's first mid-term strategy for 2011–2015 linked foreign aid with foreign policy, which was one of the six strategic priorities. The South Korean government aimed to promote ODA that would pursue national interests based on diplomatic considerations and to elevate ODA

[5] Although the initiatives were abolished after the impeachment in 2017, projects connected to the initiatives were combined with existing sectoral ODA projects.

as a core pillar of South Korea's foreign policy as a means to enhance national prestige and soft power (CIDC, 2010). However, the link was less explicit in the second mid-term strategy for 2016–2020, with only a brief mention of priority partner countries selected based on the economic and diplomatic relationship with South Korea (CIDC, 2015). Close alignment between foreign aid and foreign policy was most clearly mentioned in the 2020 action plan. The plan mandated the need to develop the third mid-term strategy for 2021–2025 based on domestic and foreign policies, key policy goals, and regional and sectoral strategies, including the New Southern Policy and the New Northern Policy. The plan also stressed the need to link the mid-term strategy, annual action plan, and the CPS more effectively as well as the need to link strategies with actual ODA projects and programs (CIDC, 2019). This implies that foreign aid is and will continue to be a crucial component of South Korea's diplomatic relations (Fig. 3).

In terms of ODA budget-wise, South Korea's aid has consistently increased in the 2000s, as Fig. 1 suggests, although it fell short of reaching the 0.20% ODA/GNI target by 2020. However, amid economic challenges and the COVID-19 global pandemic, the South Korean government announced that it will increase the ODA budget by 19% from 2020 to provide around KRW 4,079 billion (USD 3,651 million) in 2021 to help developing countries combat COVID-19 and climate change as well as to achieve the SDGs, among other goals (CIDC, 2020b).

The third Mid-Term Strategy for Development Cooperation for 2021–2025 was published in January 2021, and the strategy identifies 12 priority tasks to advance South Korea's ODA. The feature of aligning the government's overarching foreign strategy with foreign aid for synergy is explicitly stated in the strategy, including with the New Southern Policy Plus,[6] the New Northern Policy, and the summit diplomacy (CIDC, 2021).

It is along this line that Asia, the target region of both the New Southern Policy and New Northern Policy, is highly prioritized in South Korea's ODA. The government has produced a comprehensive list of priority partner countries for both grants and concessional loans since 2011 (see Table 8). While the government has maintained its focus on

[6] The New Southern Policy Plus is an upgraded version of the New Southern Policy announced in November 2020.

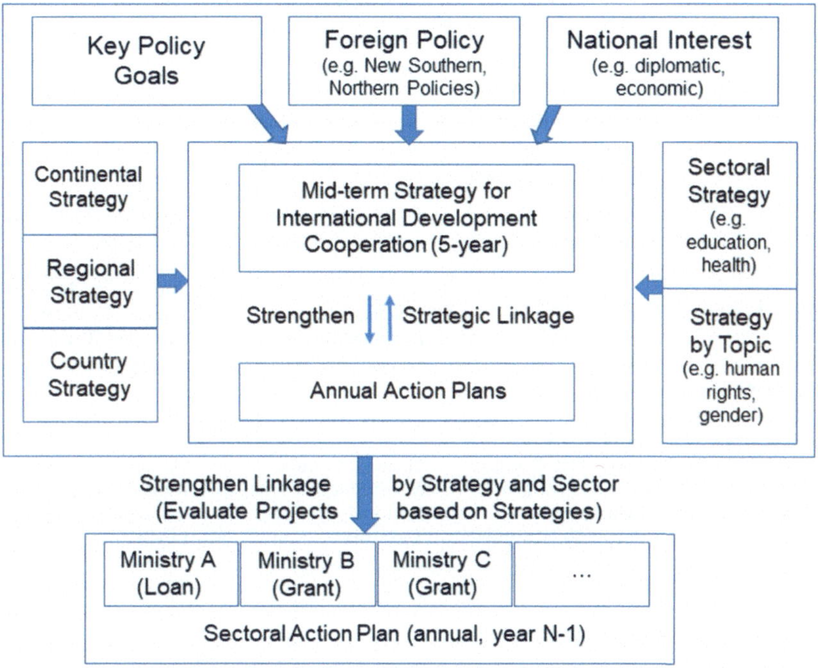

Fig. 3 South Korea's plan to strengthen strategic linkages (*Source* CIDC [2019])

Asia, the latest list of priority partner countries reflects an interconnection between foreign policy and foreign aid. India, the partner country of the New Southern Policy, and Kyrgyzstan, Ukraine, and Tajikistan, the partner countries of the New Northern Policy, were newly added to the list.

4 Conclusion

This chapter analyzed South Korea's history as an aid donor from the 1960s to the present by focusing on the overall diplomatic goals and how they were reflected in foreign aid. It also delved into the changing institutional structure and the aid disbursement pattern to identify the driving factors behind South Korea's ODA.

Table 8 South Korean ODA's priority partner countries

	2011–2015	*2016–2020*	*2021–2025*
Asia	Vietnam, Indonesia, Cambodia, Philippines, Bangladesh, Mongolia, Laos, Sri Lanka, Nepal, Pakistan, Timor-Leste	Nepal, Laos, Mongolia, Myanmar, Bangladesh, Vietnam, Sri Lanka, Indonesia, Cambodia, Pakistan, Philippines	Nepal, Laos, Mongolia, Myanmar, Bangladesh, Vietnam, Sri Lanka, Indonesia, Cambodia, Pakistan, Philippines, India
Latin America	Colombia, Peru, Bolivia, Paraguay	Bolivia, Colombia, Paraguay, Peru	Bolivia, Colombia, Paraguay, Peru
Eastern Europe•CIS	Uzbekistan, Azerbaijan	Uzbekistan, Azerbaijan	Uzbekistan, Kyrgyzstan, Ukraine, Tajikistan
Middle East	Solomon Islands		
Africa	Ghana, DR Congo, Nigeria, Ethiopia, Mozambique, Cameroon, Rwanda, Uganda	Ghana, Rwanda, Mozambique, Senegal, Ethiopia, Uganda, Tanzania	Ghana, Ethiopia, Rwanda, Uganda, Tanzania, Senegal, Egypt
Total	26	24	27

Source KGDC (2013) and ODAKorea (2020)

The findings indicate that South Korea's foreign aid was utilized as a foreign policy tool but in different ways. Until the late 1970s, its development assistance was used to support South Korea's political rivalry with North Korea in the United Nations. The underlying motivation for diplomatic engagement with the Global South, at times via foreign aid, was the fundamental competition with North Korea to secure more allies in the international political arena. However, rapid economic development and successful organization of global events, including the 1988 Seoul Olympics, led to growing demand both domestically and internationally for its responsibility to the international community as a newly industrialized country. Expanding trade relations with Asia and the need to secure support from the region also meant prioritization of Asia in terms of both foreign policy and foreign aid. However, as an emerging donor and middle power, South Korea simultaneously increased its aid allocation to Africa. To summarize, South Korea actively adapted foreign policy to the changing circumstances and utilized foreign aid as a useful tool to lead to better results and create synergies.

South Korea's aid has always been heavily influenced by economic and geopolitical factors at both the domestic and international level, albeit with different styles depending on the timeframe. While poverty, development, and global goals are at the center of its ODA, it will be nearly impossible to distinguish diplomatic interests from foreign aid. With now over 60 years of history as an aid donor, South Korea should pursue advancement in both the quantity and quality of aid. Yet, as South Korea is also economically affected by COVID-19 and is in competition with neighboring Asian donors such as China and Japan for a stronger presence, especially in Asia, it remains to be seen what its diplomatic and thereby foreign assistance strategy will lead to during these uncertain times.

References

Apodaca, C. (2017). Foreign Aid as foreign policy tool. In *Oxford Research Encyclopedia of Politics* (pp. 1–20).

Arel-Bundock, V., Atkinson, J., & Potter, R. A. (2015). The limits of foreign aid diplomacy: How bureaucratic design shapes aid distribution. *International Studies Quarterly, 59*(3), 544–556.

Atwood, J. B., Mcpherson, M. P., & Natsios, A. (2008). Arrested development: Making foreign aid a more effective tool. *Foreign Affairs, 87*(6), 123–132.

Chae, R. (2014). Diplomatic war: Inter-Korean relations in the 1970s. *Seoul Journal of Korean Studies, 27*(2), 307–330.

Cho, T., & Kim, M. (2020). *Estimating long-term economic growth in North Korea: 1956–1989.*

Choi, C.-K. (1984). The Korean Questions at the United Nations: Special Reference to the Unification Problems of Korean Peninsula. *Social Science and Policy Research, 6*(2), 55–80. (in Korean).

Chun, D. H. (1981). *Administrative Policy Speech on the 1982 Government Budget* (in Korean). http://pa.go.kr/research/contents/speech/index.jsp. Accessed 11 November 2020.

Chun, H.-S. (1996). Koreas Foreign Policy towards Southeast Asia: Possibility and limitation. *Korean Political Science Review, 30*(1), 391–414. (in Korean).

CIDC. (2010). *Mid-term strategy for development cooperation (2011–2015).* Government of Korea: Committee for International Development Cooperation (in Korean). http://odakorea.go.kr/oz.main.OdaMain.do. Accessed 11 November 2020.

CIDC. (2015). *Mid-term strategy for development cooperation (2016–2020)*. Government of Korea: Committee for International Development Cooperation (in Korean). http://odakorea.go.kr/oz.main.OdaMain.do. Accessed 11 November 2020.

CIDC. (2019). *2020 International development cooperation action plan*. Government of Korea: Committee for International Development Cooperation (in Korean). http://odakorea.go.kr/oz.main.OdaMain.do. Accessed 11 November 2020.

CIDC. (2020a). *2020 International development cooperation action plan (Final Budget)*. Government of Korea: Committee for International Development Cooperation (in Korean). http://odakorea.go.kr/oz.main.OdaMain.do. Accessed 27 April 2021.

CIDC. (2020b). *2021 International development cooperation action plan (Proposed Budget)*. Government of Korea: Committee for International Development Cooperation (in Korean). http://odakorea.go.kr/oz.main.OdaMain.do. Accessed 27 April 2021.

CIDC. (2021). *Third mid-term strategy for international development cooperation*. Government of Korea: Committee for International Development Cooperation (in Korean). http://odakorea.go.kr/oz.main.OdaMain.do. Accessed 27 April 2021.

Colombo Plan. (2020). *The Colombo Plan*. https://colombo-plan.org/history/. Accessed 11 November 2020.

GoK. (2008). *New government, 5 national visions and 21 strategies*. Government of Korea (in Korean). https://www.korea.kr/special/policyFocusView.do?newsId=148648508&pkgId=49500347. Accessed 20 January 2021.

GoK. (2011). *Lee Myung-bak administration: 100 policy tasks*. Government of Korea (in Korean). https://www.korea.kr/fcatalog/access/combo.jsp?callmode=normal&catimage=&eclang=ko&Dir=464&um=s&start=1. Accessed 27 April 2021.

GOV. (1986). *1986 Administrative White Paper*. Government of Korea (in Korean).

GOV. (2017). *2017 Korea's ODA White Paper*. Government of Korea (in Korean).

KDI. (1983). *International Development Exchange Program, 1982–83*. Korea Development Institute.

KGDC. (2013). *Improving the selection of priority partner countries and country partnership strategy—From diplomatic perspective*. Korea Global Development Consulting Center (in Korean).

Kim, D. J. (2001). *Keynotee speech at the KOR-ASEAN summit*. http://www.pa.go.kr/research/contents/speech/index.jsp. Accessed 2 November 2020.

Kim, H. S. (1990). *South Korea's mid-long term economic cooperation plan with developing countries*. Korea Institute for International Economic Policy (in Korean).
Kim, J. H. (2014). *South Korea's motivations for official development assistance: 1963–2010* (PhD thesis). Ewha Womans University.
Kim, Y. S. (1994). *Speech at the welcoming dinner with business people* (in Korean). http://www.pa.go.kr/research/contents/speech/index.jsp. Accessed 11 November 2020.
Kim, Y. S. (1995a). *Keynote speech at the UN world summit on social development* (in Korean). http://www.pa.go.kr/research/contents/speech/index.jsp. Accessed 3 November 2020.
Kim, Y. S. (1995b). *Special interview celebrating 45 years of the Korea Times* (in Korean). http://www.pa.go.kr/research/contents/speech/index.jsp. Accessed 2 November 2020.
KOICA. (1997). *1996 KOICA Annual Report*. Korea International Cooperation Agency (in Korean).
Koo, Y. (1980). Future perspectives on South Korea's foreign relations. *Asian Survey, 20*(11), 1152–1163.
Lee, C.-J. (1978). The direction of South Korea's foreign policy. *Korean Studies, 2*(1), 95–137. https://doi.org/10.1353/ks.1978.0011
MOFA. (1962). *(No. 1279) South Korea's Aid Plan to Laos*. Ministry of Foreign Affairs (in Korean). https://diplomaticarchives.mofa.go.kr/new/main/. Accessed 11 November 2020.
MOFA. (1967). *(No. 2405) South Korea's (Grant) Aid to Lesotho*. Ministry of Foreign Affairs. https://diplomaticarchives.mofa.go.kr/new/main/. Accessed 11 November 2020.
MOFA. (1968). *(No. 2811) South Korea's (Grant) Aid to Rwanda*. Ministry of Foreign Affairs (in Korean). Available at https://diplomaticarchives.mofa.go.kr/new/main/. Accessed 11 November 2020.
MOFA. (1969a). *(No. 3290) South Korea's Aid Plan for the Mekong Region of Vietnam*. Ministry of Foreign Affairs (in Korean). https://diplomaticarchives.mofa.go.kr/new/main/. Accessed 11 November 2020.
MOFA. (1969b). *1969 diplomatic records*. Ministry of Foreign Affairs (in Korean). https://diplomaticarchives.mofa.go.kr/new/main/. Accessed 11 November 2020.
MOFA. (1971). *South Korea's foreign affairs in the 1960s*. Ministry of Foreign Affairs (in Korean). https://diplomaticarchives.mofa.go.kr/new/main/. Accessed 11 November 2020.
MOFA. (1983). *Republic of Korea compilation of diplomatic documentation*. Ministry of Foreign Affairs (in Korean).

MOFA. (1987). *(No. 25681) Establishment of the EDCF, 1981–87*. Ministry of Foreign Affairs (in Korean). https://diplomaticarchives.mofa.go.kr/new/main/. Accessed 11 November 2020.

MOFA. (1989). *(No. 28870) Management of the EDCF, 1988–89*. Ministry of Foreign Affairs (in Korean). https://diplomaticarchives.mofa.go.kr/new/main/. Accessed 11 November 2020.

MOFA. (2009). *60 years of South Korea's Foreign Affairs*. Ministry of Foreign Affairs (in Korean).

MOFAT. (2006). *Korea's initiative for Africa's development*. Ministry of Foreign Affairs and Trade. https://overseas.mofa.go.kr/ch-geneva-ko/brd/m_8853/view.do?seq=589964&srchFr=&srchTo=&srchWord=&srchTp=&multi_itm_seq=0&itm_seq_1=0&itm_seq_2=0&company_cd=&company_nm=&page=50. Accessed 11 November 2020.

MOGA. (1986). *Organization of the Government of Korea*. Ministry of Government Administration (in Korean).

Morgenthau, H. (1962). A political theory of foreign aid. *The American Political Science Review, 56*(2), 301–309.

NLIC. (1986). *Foreign Economic Cooperation Fund Act*. National Law Information Center (in Korean). https://www.law.go.kr/LSW//lsInfoP.do?lsiSeq=3457&chrClsCd=010203&urlMode=engLsInfoR&viewCls=engLsInfoR#0000. Accessed 11 November 2020.

ODA Korea. (2020). *ODA priority partner countries* (in Korean). http://odakorea.go.kr/ODAPage_2018/category02/nation_stats1_nepal.jsp. Accessed 2 November 2020.

OECD. (2021). *Official development assistance 2020—Preliminary data, DAC statistics*. https://public.tableau.com/views/ODA_GNI/ODA2020?:language=fr&:display_count=y&publish=yes&:origin=viz_share_link?&:showVizHome=no#1. Accessed 20 January 2021.

OPM. (1979). *1979 Basic Management Plan*. Office of the Prime Minister (in Korean).

OPM. (1980). *1980 Basic management plan*. Office of the Prime Minister (in Korean).

Palmer, G., Wohlander, S. B., & Morgan, T. C. (2002). Give or take: Foreign aid and foreign policy substitutability. *Journal of Peace Research, 39*(1), 5–26.

Park, C. H. (1965). *1965 President's State of the Union message to the National Parliament (in Korean)*. http://pa.go.kr/research/contents/speech/index.jsp?spMode=view&catid=c_pa02062&artid=1305509. Accessed 11 November 2020

Roh, T. W. (1991). *Keynote speech at the 46th UN General Assembly* (in Korean). http://www.pa.go.kr/research/contents/speech/index.jsp. Accessed 11 November 2020.

Statistics Korea. (2021). *K-Indicator: Economic growth rate*. https://www.index.go.kr/unify/idx-info.do?idxCd=4201&clasCd=7. Accessed 25 January 2021.
Tarnoff, C., & Lawson, M. (2016). *Foreign aid: An introduction to U.S. program and policy* (pp. 1–34). Congressional Research Service Report.

PART II

Role of Private Sectors in Development Cooperation

CHAPTER 5

Quest for Sublation of Economic Development and Poverty Reduction: Dual Features of Japan's Aid in the Post-Cold War Era and After

Hyomin Jung and Motoki Takahashi

1 Introduction

The characteristics of international aid regimes have changed in accordance with the world's politico-economic trends. Following the fall of the Berlin Wall and the end of the Cold War, in the latter half of the 1990s, the majority of the member countries of the Organisation for Economic Co-operation and Development–Development Assistance Committee (OECD-DAC) and most international development

H. Jung (✉) · M. Takahashi
Graduate School of Asian and African Area studies, Kyoto University, Kyoto, Japan
e-mail: junghyominn@gmail.com

M. Takahashi
e-mail: motokitakahashi397@hotmail.com

H.-j. Kwon et al. (eds.), *International Development Cooperation of Japan and South Korea*,
https://doi.org/10.1007/978-981-16-4601-0_5

organizations, agreed to prioritize poverty reduction and *human development* (inter alia, basic education and primary healthcare) partly due to their deep concern about the situation in Least Developed Countries (LDCs), especially in Africa. Strongly influenced by this agreement, the Millennium Developments Goals (MDGs) of the 2000 UN Millennium Summit, which largely determined the future direction of international aid, also focused on poverty reduction and human development.

On the other hand, Japan had found the importance of collaboration in *industrialization* for developing countries to be one of the most important purposes of aid, especially since the middle of the 1980s when the expansion of her aid volume and success in industrialization of East Asian countries began to be apparent. In 1987, the Ministry of International Trade and Industry (MITI) stated the idea that Japan's trade-investment-aid synergetic approach had been effective for the industrialization of recipient countries (MITI, 1987). Behind this approach, there was an understanding that trade and investment with recipient countries would bring economic benefits not only to them but also to Japanese firms and the country's whole economy through exports, imports, and the securing of needed commodities, as well as dividends of investment. In the same vein, in 1992, the first Japanese ODA Charter stipulated the importance of economic factors in development among others.

Japan continues to maintain the main features of aid in support of economic infrastructure as a major target of yen-denominated development loans (hereinafter loan aid). However, after the 1990s, Japan's aid for poverty reduction in LDCs was intensified. This appears to be an effort on Japan's part to transform her aid, at least to some extent, so that it aligns more closely with the ethical requirements of the international donor community, including DAC, in terms of friendliness towards low-income countries. Moreover, in her renewed ODA Charter in 2003, Japan officially adopted the concept of *Human Security*, which literally endorses international endeavors to address the predicaments in LDCs, which include poverty, hunger, illiteracy, low levels of education and high mortality rates as well as insecurity caused by conflicts and violence. In essence, human poverty reduction and peace-building came to the fore in Japan's aid from this point on. Unfortunately, we find that these new aspects of Japan's aid have not drawn much attention or been examined in the academic literature, either in Japan or abroad.

This chapter examines how Japan's aid changed to promote poverty reduction and human development in developing countries from the late

1990s to 2000s. We should not forget that at the same time Japan has never withdrawn from the idea that aid is for economic development and the mutual benefit of both the recipient and donor. Japan's aid, especially after the 1990s, is not as simple as most Western and some domestic observers' perception. The chapter attempts to describe a firmly formulated, but yet unseen aspect of Japan's aid focusing especially on the period from the 1990s to the 2010s and thereby clarify the complex characteristics of Japan's aid.

Among the factors which caused the reorientation of Japan's aid toward poverty reduction and human development is the connection and interaction with the international *epistemic community* of development. We attempt to apply Haas's theory of epistemic community (Haas, 1992). Haas introduced the theory of epistemic community, "a network of professionals with recognized expertise and competence in a particular domain and an authoritative claim to policy-relevant knowledge within that domain or issue-area" (1992, p. 3). In this chapter, we define the international epistemic community of development as an international community of professionals which play roles to share and formulate ideas, norms, and rules of development aid to address people's poverty and insecurity. The ideas, norms, and rules are adopted by aid donors such as DAC members and international development organizations according to each donor's interests and understanding.

In this chapter, through discussion, we attempt to analyze how a donor's aid is differentiated from and homogenized with other donors. We hope it will contribute to further academic discussions on how and why there are various types of donors and aid, as well as the commonalities donors have with each other.

The chapter is organized as follows. Section two will first explain how the Japanese idea of aid differs from that of DAC Western donors, particularly those in Northwestern Europe. Section three will clarify the unseen aspects of Japan's aid, focusing on the 1990s to 2000s. In Section four, the epistemic community of development and its influences on and interaction with Japanese professionals will be discussed to explain the changing features of Japan's aid since the 1990s. Finally, we will summarize and present the changing nature of Japan's aid in comparison with other donors in the 2010s as a conclusion. Japan's overlooked but continued efforts with regards to poverty reduction as well as reemphasis of economic development and situation under circumstances of the COVID 19 pandemic will be briefly discussed.

2 Japanese Perceptions on Aid Before the 1990s: Views Different from the International Mainstream

Japan's aid has long been characterized as an investment in large shares of economic infrastructure through loan aid with East Asia as its regional target. Even though efforts were made to "improve the quality" of Japan's aid in the 1980s, the perception that Japan's aid was based on a strong commercial interest did not disappear (Arase, 1995; Mawdsley, 2012).

There was a large gap in the perception of aid effectiveness, especially in the early 1990s, between Japan and other DAC donors. Conceptions of aid effectiveness shared by the majority of DAC members, except for Japan, were full of disappointment in the early 1990s. Many DAC donor officials and the World Bank seriously doubted whether aid really worked in developing countries, notably in LDCs. Despite the aid given by Western European countries to LDCs, the majority going to Sub-Saharan Africa, economic stagnation continued. It seemed to them, that investment in economic infrastructure supported by loan aid from donors merely led to huge debt burdens for people in LDCs and thereby hindered poverty reduction. Moreover, Structural Adjustment Programs (SAPs) led by the World Bank and International Monetary Fund (IMF) from the 1980s and on could not achieve their intended results (Easterly, 2006). Also, the programs were criticized by UNICEF and other donors as aggravating rather than alleviating human poverty (Cornia et al., 1987). Compounded with the drastic changes of the global strategic situation at the end of the Cold War, many donor governments questioned whether spending taxpayer money for aid was meaningful. *Aid fatigue*, the decline in motivation for giving aid was an inevitable consequence.

Despite the aid fatigue in the West, Japan continued to expand her aid. As a consequence, Japan became the largest aid donor, except 1990, from 1989 to 2000. As Western countries restrained their aid, Japan came to the fore. Positive views on aid effectiveness were widely shared by the country's bureaucracy and the conservative wings of media and academia. However, the liberalist media and academia were critical of the negative impacts of ODA such as reinforcement of authoritarian regimes, corruption, human rights violations, and environmental damage.

In East Asia, newly industrializing economies (NIEs), namely South Korea, Taiwan, Hong Kong, and Singapore had succeeded in rapid economic growth and Southeast Asian nations had started to follow suit. Most of those nations had been the main recipient countries of Japan's aid. In *The East Asian Miracle* published in 1993, these countries and their economies were designated High Performing Asian Economies (HPAEs) by the World Bank. HPAEs include Japan, the four Asian NIEs and other major recipients of Japan's ODA such as Thailand, Malaysia, and Indonesia (Takahashi, 2015; World Bank, 1993).

The positive views of aid by Japanese bureaucrats before the 1990s, had two major streams, represented by the Ministry of Foreign Affairs (MOFA) and MITI. Officials of the Economic Cooperation Bureau of MOFA published a report, titled *The Ideas of Economic Cooperation*, to explain why Japan should provide aid for developing countries and cited humanitarian consideration and importance of mutual dependence in the international community as justifications for ODA (MOFA, 1981). This was the diplomats' attempt to introduce the internationally accepted common-sense ideas of aid. Humanitarian consideration and the importance of mutual dependence have been repeated as the most fundamental justification for aid in MOFA's annual white papers and ODA charters. It is important to note that the MOFA bureau responsible for ODA was indeed titled the *Economic* Cooperation Bureau, however.

The second stream of positive views of aid was typically put forward by MITI. MITI officials thought that Japan's aid, through trade-investment-aid synergy, had been a great benefit to the economic development of Asian countries (MITI, 1987). This view was shared by those in the Overseas Economic Cooperation Fund (OECF), the agency in charge of implementing loan aid, and the Ministry of Finance (MOF) which supervised loan aid and thus the OECF. Here the term, economic development, meant more or less industrialization (Kohama & Teranishi, 1992, p. 16). One of the origins of this identification of economic development with industrialization was the Fukuda doctrine (1977), found in the premier's speech vowing a substantial change in Japanese diplomatic policies vis-à-vis Southeast Asian nations, which stressed that Japan should support the recipients' industrialization.[1]

[1] The Fukuda doctrine was presented in Manila, the Philippines by then Prime Minister Takeo Fukuda to restore relations with Southeast Asian countries which had been seriously damaged partly due to their perception of Japan's selfish attitude toward them. Fukuda, in his speech, wanted Southeast Asians to know that, "Japan has already announced a policy of more than doubling its official development assistance within the next five years. We anticipate that an important part of this assistance will continue to be for

Thanks in part to the large amounts of aid, Japan's aid actors became more self-assertive in the early 1990s. In 1991, the OECF published a paper, "Issues related to the World Banks approach to structural adjustment: a proposal from a major partner" (OECF, 1991). The OECF provided approximately 450 billion yen as co-financing for the SAPs from 1986 to 1991 (MOFA, 2004). It argued that the Bank's approach should be reconsidered as the OECF thought that it overly prioritized the roles of market mechanisms and downplayed those of government. The OECF repeated that it was valuable to learn from Japan's experiences, having had success when the government purposefully intervened in markets for the promotion of development.

Also, the MOF, which is in charge of contributions to and dialogue with the World Bank, supported the publishing of *The East Asian Miracle* by the bank, which positively described HPAEs and analyzed the mechanisms of the East Asian countries' successful experiences of economic development (Suzuki, 2018, p. 15; World Bank, 1993). In the decade following, dissemination and transfer of "Japan's (and Asia's) experience" became the main agenda of the country's aid.

The first ODA charter in 1992 was the Japanese government's comprehensive response to the criticism that it had a lack of ideas. The charter emphasized the importance of economic development as holistic development and further added that "Japan attaches central importance to the support for the *self-help efforts* of developing countries towards economic take-off." In addition, the ODA charter emphasized the successful economic development Japan and other successful Asian countries had experienced (MOFA, 1992).

The idea of support for self-help efforts was included in the ODA charter in addition to humanitarian considerations and the importance of mutual dependence. Many features were pointed out as elements of the self-help efforts (Takahashi, 1998; Watanabe, 1991). One element, the *request-based* aid principle stated in the 1992 ODA charter, meant that aid was to be provided at the initiative of the recipient countries (MOFA, 1992). Another element was the recipient countries' responsibility to bear the burden of paying for portions of the resources for development. It is believed that recipient countries' ownership would be respected and maximized with conditions such as the recipients' responsibilities to repay

industrial projects, or for infrastructure improvement which will facilitate industrialization in Southeast Asia" (Fukuda doctrine speech, 1977).

the loans (Nishigaki et al., 2009). This is based on the assumption that the recipient countries have the management ability and commitment to implement development policies/projects supported by aid as proposed.

The self-help effort notion was indirectly related to interactions between Japan and the international aid communities' perceptions of aid. Self-help had become the core principle of Japan's aid as stated in the ODA charter, and Japan thought this idea could be translated as "ownership" in an important strategic document titled the "Shaping the 21st-century" adopted as the consensus of OECD-DAC (JICA, 1998; OECD, 1996a). However, agreement among Western DAC member countries was based on their past disregard of the motivations of recipients and the demanding of SAP conditionality unilaterally, which largely failed and apparently caused a deepening of poverty. Japan did not share the Westerners' grave reflection on this point.

Meanwhile, "Shaping the twenty-first century," newly prioritized human development such as education and health and became one of the precursory documents that developed into the MDGs. "Ownership" subsequently became one of the five principles in the Paris Declaration on Aid Effectiveness in 2005, another important document representing the efforts to improve aid effectiveness through coordination among aid stakeholders such as donors and recipients to achieve the MDGs.[2]

Taking advantage of her position as the largest donor among the DAC members in the 1990s, Japan tried to lead the donor communities by suggesting directions and putting emphasis on economic development and self-help efforts. Unfortunately, its persuasive efforts faltered after the Asian financial crisis of 1997. Most HPAEs suffered negative impacts from the crisis. The financial crisis resulted in the weakening of Japan's claims of superiority for her own development policies and approaches to aid. In addition to this, the volume of Japan's aid had already started to decrease after the mid-1990s due to Japan's acute fiscal situation and East Asians' graduation from the status of aid recipients.

However, after the latter half of the 1990s poverty reduction and peace-building increasingly became the main concerns for not only the international community but also for Japan (Takahashi, 2018). There were continuous efforts to reinforce the idea of assistance for human development including basic education and primary healthcare and

[2] The five principles of the Paris Declaration are ownership, alignment, harmonization, managing for results, and mutual accountability (OECD, 2005).

humanitarian relief for victims of armed conflicts as parts of Japan's aid. In an effort to take part in the international paradigm, human security was heralded as a part of Japan's aid since the end of the 1990s. Prime Minister Keizo Obuchi stated Japan's commitment to human security as part of a countermeasure to the Asian economic crisis in a speech in Vietnam in 1998.[3] Human security became one of the main ideas in the revised ODA Charter in 2003, and was succeeded by the Development Cooperation Charter, revised in 2015.

As a result, over time, in the 1990s and 2000s, the ideas and modalities of Japan's aid acquired dual characteristics. Japan's aid, especially in the mid-1990s to 2000s, on the whole, substantially maintained features formulated before the 1990s such as large shares of loan aid as modality and economic infrastructure as the target sector. On the other hand, efforts toward poverty reduction and human development newly obtained increasing priority in Japan's aid, especially in regards to LDCs. This change has received little academic attention even though its meaning and impact on the size and distribution of aid and aid activities has been considerable. In the next section, we do our best to clarify the details of these overlooked aspects of Japanese aid from the 1990s to 2000s.

3 Overlooked Aspects of Japan's Aid After the Late 1990s and 2000s

National leaders all over the world including Japan and other DAC member countries and international organizations reached an agreement in 2000, represented by MDGs focusing on human poverty reduction (education and healthcare) and improvements of social welfare in developing countries. It is no wonder then that the majority of DAC donors and international organizations, notably those under the auspices of the UN, made much of the support for the human development sectors such as education and health through grants, especially for LDCs.

It is important to recognize that Japan tried to promote poverty reduction and human development equally to that of other DAC donors, though at the same time she did not change her tendency to also emphasize support for economic development. "Poverty reduction through economic growth" was put forward by MOFA as an important slogan

[3] Keizo Obuchi's speech in Vietnam (1998).

Table 1 Share of grants (including technical assistance) and loans in Japan's aid (Gross Disbursements)

	1980	*1985*	*1990*	*1995*	*2000*	*2005*	*2010*	*2015*	*2019*
Grants	1,525.2	2,493.2	3,609.4	4,701.8	4,991.3	8,776.8	5,573.5	5,486.1	5,181.9
	30.3%	38.8%	36.8%	47.0%	45.2%	57.8%	45.8%	41.8%	36.0%
Gross	3,502.9	3,936.7	6,186.4	5,312.1	6,049.3	6,415.4	6,584.4	7,646.2	9,227.6
Loans	69.7%	61.2%	63.2%	53.0%	54.8%	42.2%	54.2%	58.2%	64.0%
Total	5,028.1	6,429.9	9,795.9	10,013.9	11,040.6	15,192.2	12,157.9	13,132.3	14,409.5
	100%	100%	100%	100%	100%	100%	100%	100%	100%

Sources OECD stat (US million dollars, %). Accessed on March 8, 2021 from https://stats.oecd.org/#

in its ODA white papers in the 2000s (cf. MOFA, 2005). It stressed the importance of economic development on the basis of Japan's experiences in East Asia. On the other hand, Japan also agreed to the MDGs and the 2005 Paris Declaration on Aid Effectiveness, harmonizing with other donors and doing what it could to contribute to achievement of these goals.

According to Table 1, the share of grants increased in the late 1990s and 2000s compared to the share of loans in the 1980s and 2010s. Grants accounted for 30.3% and 38.8% in 1980 and 1985 respectively. The grant rate (including technical cooperation) increased to 47.0% in 1995, and was at a record high of 57.8% in 2005.

In addition to the increase in the grant rate, Table 2 shows the distribution of Japan's loan aid by development and income-level classification group in the 1990s and 2000s compared to other DAC countries' totals in the same period. Japan's loan aid to LDCs was recorded as low as 3.8% and 2.5% in the 1990s and 2000s respectively. The shares of other DAC countries' sums were larger, recorded at 11.4% and 11.0% in the 1990s and 2000s respectively. This means that Japan's loan aid to LDCs was more concessional with less burden for the recipients.[4]

Japan's share of grant allocations to LDCs does not show a huge difference in comparison with the average of other DAC countries' distribution by development and income level classification group as shown in Table 3. In the 1990s and the 2000s, DAC's average share of grant aid to LDCs

[4] It means that Japan's loan aid was more concentrated on the Low and Middle-income Countries (LMICs) and Upper Middle Income Countries (UMICs).

Table 2 Japan and other DAC countries loan distribution by recipients' development/income level (Gross Disbursements)

	Japan 1990–1999	*DAC countries' sum except Japan 1990–1999*	*Japan 2000–2009*	*DAC countries' sum except Japan 2000–2009*
LDCs, Total	2,507.4 3.8%	6,624.4 11.4%	1,569.6 2.5%	4,110.8 11.0%
Other LICs, Total	85.2 0.1%	549.7 0.9%	52.9 0.1%	9.0 0.0%
LMICs, Total	39,023.6 59.8%	32,156.7 55.3%	36,840.4 59.2%	18,476.4 49.3%
UMICs, Total	22,216.3 34.0%	15,686.4 27.0%	23,145.6 37.2%	11,156.7 29.8%
MADCTs, Total	1,404.1 2.2%	2,456.0 4.2%	26.5 0.0%	331.5 0.9%
Part I Unallocated by income	43.6 0.1%	685.1 1.2%	558.0 0.9%	3384.0 9.0%
Total	65,280.1 100%	58,158.3 100%	62,193.0 100%	37,468.3 100%

Note MADCT stands for "More Advanced Developed Countries and Territories"
Sources OECD stat (US million dollars, %). Accessed on March 8, 2021 from https://stats.oecd.org/#

was 27.8% and 28.2% respectively, compared to Japan's 24.1% and 25.7% respectively, which are slightly lower than the share of the average of DAC countries except for Japan, the differences being only 3.7% point and 2.5% point respectively. The shares of grants to LMICs were also not largely different between other DAC countries and Japan. On the other hand, Japan's grants to UMICs were 20.5% and 25.8% in the 1990s and 2000s respectively, unlike other DAC countries at 12.9% and 19.8%.

From Tables 2 and 3, it is clear that Japan considered the varied developmental and financial conditions of recipient countries since Japan distributed only a small share of its loan to LDCs in comparison with other DAC countries though the total absolute size of the loan was huge. In addition, Japan's grant share to LDCs and LMICs was slightly lower than other DAC countries, but the difference was not outstanding. Japan seemingly consciously allocated more loans to countries with higher incomes, and grants to lower-income countries at almost the same rate as other DAC countries. This implies that Japan's aid focused not only

Table 3 Japan and other DAC countries grant distribution by recipients' development/income level

	Japan 1990–1999	*DAC countries' sum except Japan 1990–1999*	*Japan 2000–2009*	*DAC countries' sum except Japan 2000–2009*
LDCs, Total	11,323.5 24.1%	85,460.2 27.8%	16,043.2 25.7%	163,249.7 28.2%
Other LICs, Total	365.7 0.8%	2,655.4 0.9%	115.4 0.2%	3,413.2 0.6%
LMICs, Total	13,568.5 28.8%	79,851.6 26.0%	16,264.9 26.1%	137,714.9 23.8%
UMICs, Total	9,636.0 20.5%	39,573.0 12.9%	16,115.3 25.8%	11,4815.7 19.8%
MADCTs, Total	1,541.8 3.3%	22,839.3 7.4%	344.4 0.6%	4,503.6 0.8%
Part I Unallocated by income	10,623.0 22.6%	77,097.3 25.1%	13,545.7 21.7%	154,807.6 26.8%
Total	47,058.5 100%	307,477.0 100%	62,428.9 100%	578,504.6 100%

Sources OECD stat (US million dollars, %). Accessed on March 8, 2021 from https://stats.oecd.org/#

on loans for economic development but also on social infrastructure and human development in the LDCs.

Strong requests for debt relief from donors forced Japan to offer relief from her official lending to the scheme of forgiving Heavily-Indebted Poor Countries (HIPCs) in the 2000s. Debt cancellation came to account for a large part of her ODA. Japan committed to actions relating to debt relief, mainly in the mid-2000s. In 2003, debt relief accounted for 43.1% of her bilateral aid so economic infrastructure hit bottom, recorded at 21.7% the lowest throughout all the years of Japan's ODA.[5] The share of "social infrastructure and services" also increased in the 2000s, perhaps an indication of Japan's commitment to the MDGs.

Indonesia and Tanzania can be seen as representative examples of the dual features of Japan's aid. Indonesia can represent Japan's aid to Asia in contrast with that of Tanzania as an example of her efforts toward MDGs and poverty reduction, especially in the 1990s and 2000s. Indonesia has been one of the Japan's main recipients of aid for a long time. By 2016,

[5] Accessed on March 8, 2021 from https://stats.oecd.org/#.

Table 4 Japan's aid to Indonesia and Tanzania (Total Disbursements) (Sum of each sector from 2005 to 2010)

	Indonesia	*Tanzania*
Social Infrastructure and Services	1112.8 13.7%	167.6 12.2%
Economic Infrastructure and Services	2946.2 36.3%	378.0 27.6%
Production Sectors	543.1 6.7%	44.5 3.3%
Multisector	1521.2 18.7%	7.1 0.5%
Programme Assistance	1212.9 14.9%	109.1 8.0%
Action Relating to debt	469.5 5.8%	641.8 46.9%
Humanitarian Aid	267.4 3.3%	10.3 0.8%
Unallocated/ Unspecified	50.7 0.6%	10.8 0.8%
Total	8,123.7 100%	1369.1 100.0%

Sources OECD stat (US million dollars, %). Accessed March 8, 2021 from https://stats.oecd.org/#

Indonesia had received a total of more than 49.5 billion dollars in aid from Japan (JICA, 2018). The modality of the donor's aid to Indonesia lies mainly in loans, while numerous technical cooperation and grant projects have also been implemented. Indonesia is usually included in LMICs, the main category of countries receiving Japan's loan aid (JICA, 2018).[6] "Economic infrastructure and services" was the sector which received the largest percentage (36.3%) of the aid given to Indonesia from 2005 to 2010 (Table 4).

On the other hand, Tanzania is representative of Japan's aid to LDCs. In 2000, her debt forgiveness accounted for 70.5%. The large amount of "debt forgiveness" continued mainly from the mid-1990s to mid-2000s, showing a total share of 74.1% in 2001, 70.9% in 2004, and as large as 88.9% in 2007.[7] Grants (including debt forgiveness) were the main

[6] Indonesia was considered one of the Lower Middle Income Countries (LMICs) in 1996, but from 1997 to 2004 it was categorized as Other Low Income Countries (LICs) affected by the Asia Financial Crisis. Indonesia was categorized as a LMIC again from 2005 to 2010 (OECD, 1996b, 2000, 2001, 2003, 2006b, 2007, 2008, 2010a).

[7] Accessed on March 8, 2021 from https://stats.oecd.org/#.

Table 5 Japan's Grant and Loan Aid (Gross Disbursements) to Indonesia and Tanzania

	Indonesia		*Tanzania*	
	the 1990s	*the 2000s*	*the 1990s*	*the 2000s*
Grant	2,265.6 15.7%	1,653.4 15.7%	875.2 99.6%	1,698.0 95.0%
Loan	12,148.9 84.3%	8,878.4 84.3%	3.1 0.4%	88.8 5.0%
Total	14,414.5 100%	10,531.8 100%	878.3 100%	1,786.8 100%

Sources OECD stat (US million dollars, %). Accessed on April 27, 2021 from https://stats.oecd.org/#. The 1990s refers to the sums from 1990 to 1999, and the 2000s from 2000 to 2009

features of aid in the 1990s and 2000s. From 2005 to 2010, debt forgiveness and social infrastructure and services, accounted for 59.1% of the total in Tanzania. Given this situation there was no room for loan aid to make up a large proportion of Japan's aid to LDCs.

Japan's aid to Indonesia and Tanzania in the 1990s and 2000s consisted of substantially opposite shares of grants and loans. As Table 5 shows, loan aid to Indonesia accounted for 84.3% in the 1990s and 2000s likewise while in Tanzania it was only 0.4% and 5.0% respectively.

4 Epistemic Community and Japanese Engagements in Poverty Reduction

4.1 *Epistemic Community of International development*

The epistemic community is a well-known concept to explain how a group of experts shares beliefs and notions within a community that spreads across borders and how it affects policy-making (Haas, 1992). Application of the concept of epistemic community is an effective and necessary way to clarify the background and the Japanese commitment to and understanding of human development and poverty reduction that has not been brought to light so far. International organizations and elements of international development have also been referred to as epistemic communities in previous literature (Hopkins, 1992; Karns et al., 2004).

In this chapter, the epistemic community for international development which shares knowledge and beliefs refers to the group of connected individuals that extends among international organizations such as UN organizations, the OECD-DAC secretariat, the World Bank, research institutions, and finally each donor country's aid agencies and related organizations and experts. Of special interest is how the beliefs and notions of aid for poverty reduction and human development were shared inside the community from the 1990s to the 2000s.

According to Haas, there are four main theoretical conditions for an epistemic community to function (1992, p. 3). The first condition is "a shared set of normative and principled beliefs, which provide a value-based rationale for the social action of community members" (Haas, 1992, p. 3). Although the OECD-DAC set out the purpose of aid to be shared by member countries in the 1960s, it did not specify goals but rather conclusively set the aims of aid as broad as the recipient countries' economic development and social welfare (OECD, 2021). In addition to the two aims, international organizations and DAC member countries share beliefs such as importance of human rights and environmental issues. In addition, the DAC offers guidelines for member nations to implement their aid programs and projects (see OECD, 2006a, 2010b, 2014).

"Shared causal beliefs, which are derived from their analysis of practices leading or contributing to a central set of problems in their domain and which then serve as the basis for elucidating the multiple linkages between possible policy actions and desired outcomes" is the second condition for epistemic community (Haas, 1992, p. 3). As previously explained, since the late 1990s, members of the epistemic community of international development: the professionals in international organizations, academia, and aid agencies vowed to pursue policies and actions for poverty reduction and human development. Among them, there was broad agreement that human factors such as knowledge, to be cultivated by education, and health, to be promoted by health and medical care are likewise central in poverty reduction and development. Discussions on the problems and solutions of these global issues resulted in the setting of global aims for international development in the September 2000 MDGs.

Thirdly, "shared notions of validity – that is, intersubjective, internally defined criteria for weighing and validating knowledge in the domain of their expertise" (Haas, 1992, p. 3), can be seen within the epistemic community of international development as well. The epistemic community of international development shares the belief that evaluation of aid projects and development policies and actions is essential. The concepts and criteria of evaluation were strengthened particularly in the 2000s with the setting of MDGs. Symbolic of this point was the setting of numerical targets for education and healthcare and obliging evaluation and review of achievements in the MDGs. However, one of the main issues among the epistemic community today is still agreement on how to validate development.

Finally, "a common policy enterprise – that is, a set of common practices associated with a set of problems to which their professional competence is directed, presumably out of the conviction that human welfare will be enhanced as a consequence," is also a critical point of epistemic communities in general (Haas, 1992, p. 3) and of international development in particular. As the ultimate goal of aid, which the epistemic community pursues, is poverty reduction and human development in *recipient countries*, as outlined in the MDGs, as well as economic development and welfare as confirmed by OECD (2021). These basic concepts are shared by the professionals within the epistemic community of international development. Also, in most cases, policy enterprise for aid needs people with professional knowledge especially in the aid community.

4.2 *Japanese Engagement in Epistemic Community*

Apart from Japan's indifference until the early 1990s, symbolized by weak interest in the issue of aid effectiveness, the main focus of ODA in the international aid community shifted to poverty reduction rather than economic development from the late 1990s on. The UK under the Labour Party government and Northern European donors, the so-called like-minded group put emphasis on poverty reduction centered on basic education and primary healthcare in their ODA. The DAC secretariats, as one of the cores of the epistemic community, also played a role in the process of policy-making and implementation.

Among the advocates for debt relief were nations of the like-minded group led by the UK, which did not provide loan aid. It was agreed that it would be a requirement for a HIPC to formulate a Poverty Reduction Strategy (PRS) which outlines comprehensive policies for reform, development, and poverty reduction to be relieved of their debts, partly due to a proposal from the UK (Short, 2004, pp. 83–84). Japan was reluctant to join the plan for debt relief since its absolute and relative amounts of debt were larger than other donors, and the idea of debt relief was not compatible with the idea of self-help that Japan supported.

However, debt relief for HIPCs was eventually accepted by Japan against the backdrop of irresistible pressure from other major donors and global civil societies represented by the Jubilee 2000 movement (Short, 2004, pp. 83–84). In this instance, one could not say that Japanese government officials were active participants in the epistemic community of international development.

However, this is not to say that Japan was isolated from the epistemic community of international development. As Japan's domestic perceptions on international development progressed, Japan's aid came to show dualist features. Notwithstanding the beliefs of officials and decision makers in the superiority of East Asian development based on industrialization and the approach based on trade-investment-aid synergy, some professionals in Japan's aid community, especially in the Japan International Cooperation Agency (JICA), were connected with or participated in the epistemic community. Indeed, some officials in MOFA, largely assisted by JICA staff members and other professionals connected with the community, also began to understand the important agenda of the epistemic community (Takahashi, 2018).[8] Involvement in the epistemic community of international development had much to do with the personal background and experience of professionals that had studied in Western countries, especially the UK or the US, or had work experience in fields other than East Asia, like Sub-Saharan Africa where so many Western donors were engaged in aid and interacted with each other.[9] As a result, the epistemic community of international development steadily came to influence the Japanese perception of aid.

Commitment to poverty reduction and human development was not readily accepted by ministry officials as their appreciation of the importance of economic development was already well-rooted. Nevertheless, efforts on behalf of poverty reduction and MDGs slowly gained credence as an important aspect of Japan's aid.

The slogan "poverty reduction through economic growth" may be understood as a compromise between engagement in poverty reduction and commitment to economic development. Japan never abandoned the

[8] The authors' interviews with aid officials from 2015 to 2021.

[9] Interview with an anonymous Japanese scholar with field experience in international organizations (conducted on 18 April, 2021).

ideas of development through aid gained from her experience of cooperation leading to East Asia's industrialization but at the same time Japan also tried to adopt the ideas, then mainstream, of international aid after the late 1990s which focused on poverty reduction.

Notable and symbolic of the change in perception of Japan's aid at this time was the appointment of Sadako Ogata as president of JICA in 2003. Ogata obtained a doctoral degree in Political Science from the University of California, Berkeley, a top-level university in the US, in 1963. She taught in Japanese universities and then engaged in Japan's UN diplomacy. From 1990 to 2000, Ogata was the head of the United Nations High Commissioner for Refugees (UNHCR) and tackled refugee predicaments due to conflicts caused by socio-political destabilization after the collapse of the Cold War regime. Subsequently she became co-chair, with Amartya Sen, a Nobel laureate, of the UN Human Security Commission. She was so internationally respected that she could intellectually contribute to the epistemic community of international development by the early 2000s.

In Japan, people recognized Ogata's academic and professional work in the epistemic community and a largest number of the employees of JICA voted for Ogata as the best candidate for the future JICA president. In fact, the labor union of JICA directly asked her to assume the presidency, an extremely rare case for Japanese government agencies. The decision of the government to appoint Ogata as JICA president followed suit (Nobayashi & Naya, eds., 2020). Based on her experience in working for refugees in African and other nations, she played an important role in helping to make the Japanese version of human security to be extended to incorporate freedom from fear i.e. security in relation to violence and conflicts, in addition to freedom from want i.e. development and poverty reduction. This conception of human security was added to the aid ideas of the second ODA charter in 2003.

Thus, Japan's aid was transformed in various aspects from the 1990s to 2000s. Japan strived to foster professionals in international development cooperation as the aid volume increased. Institutions, such as graduate schools with an aim at cultivating expertise in international cooperation and development were established in many universities in the 1990s. Also, JICA and OECF, which in 1999 merged into the Japan Bank for International Cooperation (JBIC), seconded bright young staff members to study at graduate schools in the US, the UK, and other developed countries. In addition, as the public's attention increased, JICA's popularity among new university graduates improved. Thanks to these moves, the quality of the human resources, with substantial education and language proficiency, in aid agencies, including JICA, OECF, and development consulting firms, substantially increased.

By the 1990s Japan wanted to strengthen her international presence in the development sphere. For instance, in 1993, Japan sponsored the Tokyo International Conference on African Development (TICAD), which continued every five years and then every three years after 2013, which publicized Japan's commitment to Africa. The sponsorship of TICAD necessitated officials in MOFA and JICA to study and learn about international discussions regarding poverty reduction, development, and aid.

Moreover, in the 2000s Japan made efforts to participate and sometimes take the lead in aid coordination through technical assistance, non-project aid, and budget support. It was inevitable to take part in donor meetings being one of the DAC donor countries. In the case of Vietnam, the volume of aid was relatively huge, and Japan sometimes chaired the donor meetings (MOFA, 2006). Japan took the lead in donor coordination for specific sectors, such as agricultural development in Tanzania. Since 2003, the ODA Taskforce in each recipient country was organized by the Japanese embassy, field offices of JICA, and JBIC, to discuss the Country Partnership Strategy (CPS), partly based on dialogues with local communities and NGO members including advocates for poverty reduction (MOFA, 2010). Japan's aid made much progress in terms of human expertise as well as efforts to engage in poverty reduction through these activities in recipient countries' local participation.

4.3 *From Aid to Development Cooperation: Global Transmutation Through the 2010s*

From the late-2000s, in response to the relatively high growth of developing countries, especially those in Africa, and the increasing presence of emerging donors, the main focus of global aid was once again in flux. National leaders began to shift their focus of international development cooperation from human poverty reduction to other issues, including industrialization.

The common perception in the late-2000s was that development cooperation should incorporate all stakeholders, including emerging donors and recipient countries. A resolution titled "Busan Partnership for Effective Development Cooperation" represented the direction of the new change. The resolution highlighted "a set of common principles for all development actors that are key to making development co-operation effective" (OECD, 2012). Ownership, focusing on results, partnerships, transparency, and shared responsibility were the main common principles of the Busan Partnership (OECD, 2012).

At the 70th UN General Assembly in 2015, it was determined that the Sustainable Development Goals (SDGs) were to be achieved by 2030 by having 'no one left behind.' The SDGs focus on multi-dimensional issues of development, including economic growth, were very different from the MDGs. Members of the epistemic community of international development, including those in Japan, perhaps forced by these adverse circumstantial changes, also gradually ceased to be vocal about the need for poverty reduction and human development.

However, human development still remained important even with this change in focus. In the 2010s, JICA focused on quality improvement in basic education as basic education quantitatively had expanded due to the goal of education for all. For example, the "Program for Enhancing Quality of Junior Secondary Education," was implemented in Indonesia from 2009 to 2013. Japan dispatched experts, and also conducted training in Japan, as well as providing equipment (JICA, 2013, p. 9). JICA also continued to make efforts to foster human resources in the health field. In Ghana, the "Noguchi Memorial Institute for Medical Research," constructed with JICA's support in 1979, played a key role in Japan's support by making it a regional hub for the extension of health technology (JICA, 2016). The dual features of Japan's aid had seemingly taken root.

5 Concluding Remarks

Up until the 1980s, Japan's aid for developing countries focused on constructing economic infrastructure through loan aid and concentrated on Asian countries which were not necessarily LDCs. From the latter half of the 1990s to the 2000s, poverty reduction and aid for human development as proposed by the MDGs, became the main agenda of the international aid community. Partly due to differences in perceiving aid failures of others and recognizing the effectiveness of her own past experiences, Japan continued to emphasize the importance of economic infrastructure through loan aid. At the same time Japan, in line with international efforts, simultaneously began to contribute to poverty reduction and human development. For example, an emphasis on human security, including poverty reduction and peace-building, became the core idea of Japan's aid in 2003, and this continued under the 2015 Development Cooperation Charter. A number of Japanese professionals and scholars engaged with the global epistemic community of international development and influenced policy-making in Japan's aid in the context of MDGs and human development.

However, from the late-2000s, the influence of emerging donors on low-income countries increased. In this regard, in the early 2010s, the international consensus of development cooperation began to drift away from its orientation toward poverty reduction. The SDGs focused not only on poverty reduction but economic and industrial development as well. Developed countries, including major donors in DAC, also started to make their aid more in line with their national interest than in the 2000s. Japan was no exception. Since 1997, Japan had suffered from persistent deflation, and her economic situation was exacerbated in the global financial crisis of 2008, so the volume of her aid did not increase. In addition, the Great East Japan Earthquake in 2011 became another factor restraining the scale of Japan's aid. These factors accelerated the shift of Japan's aid back to a reprioritization of industrialization. There was a revision of the ODA charter in 2015 under Prime Minister Shinzo Abe making it the "Development Cooperation Charter." The goal of Japanese "national interests" was more strongly endorsed in the new charter than in previous charters.

While the SDGs diluted the focus on poverty reduction and human development, they also expanded the number of stakeholders of international cooperation and included different actors. Those in the epistemic community of international development can take advantage of the newly formulated and extended commonality of interests in international development across borders and work together again to prevent narrow-minded national interest orientation.

The world was shaken by the COVID-19 pandemic in 2020. Countries abruptly closed their borders, which made international development cooperation more difficult. JICA had no choice but to once call back officials, experts, and volunteers seconded to developing countries. Moreover, the COVID-19 pandemic seemingly accelerated countries' inward-orientation and pursuit of narrow-minded national interests.

Nevertheless, the global pandemic has also shown humankind that they cannot survive without mutual cooperation. A type of virus spread in one country can easily cross borders and threaten others, not just geographical neighbors. After the development of COVID-19 vaccines, there is an imminent need to disseminate them to vulnerable LDCs suffering the pandemic as well as the developed world. International cooperation, especially through ODA from DAC donor countries will be essential. Humankind cannot enjoy sustainable development and peace of mind and health without this. What is hoped for would be a sublation of the two orientations, i.e., a focus limited to human development and a concentration of interest in economic development. The two need to be synthesized in a complementary way and they are both fortunately incorporated as major pillars in SDGs. Strengthening international unity and collaboration to rebuild aid by synthesizing economic development and human poverty reduction is the most necessary task in this era of COVID-19 and beyond.

References

Arase, D. (1995). *Buying power: The political economy of Japan's foreign aid.* Lynne Rienner Publishers.

Cornia, G. A., Jolly, R., Stewart, F., & UNICEF. (1987). *Adjustment with a human face: Protecting the vulnerable and promoting growth: A study by UNICEF.* Oxford University Press.

Easterly, W. R. (2006). *The white man's burden: Why the West's efforts to aid the rest have done so much ill and so little good.* Penguin.

Fukuda, T. (1977, August 18). *Speech by Prime Minister Takeo Fukuda at Manila.* https://worldjpn.grips.ac.jp/documents/texts/docs/19770818.S1E.html. Accessed 2 February 2021.

Haas, P. M. (1992). Introduction: Epistemic communities and international policy coordination. *International Organization, 46*(1), 1–35.

Hopkins, R. F. (1992). Reform in the international food aid regime: The role of consensual knowledge. *International Organization, 46*(1), 225–264.

JICA. (1998). *DAC Shin kaihatsu senryaku enjo kenkyukai hokokusyo* [In Japanese: New Development Strategy Assistance Study Group Report Vol. 1]. JICA.

JICA. (2013). *Program for Enhancing Quality of Junior Secondary Education: Summary of Terminal Evaluation.* JICA Indonesia Office.

JICA. (2016). *The Project for the Construction of Advanced Research Center for Infectious Diseases at Noguchi Memorial Institute for Medical Research.* JICA.

JICA. (2018). *Indonesia ni taisuru nihon no kyouryoku no ashiato* [In Japanese: Footprints of Japan's cooperation with Indonesia]. https://www.jica.go.jp/publication/pamph/region/ku57pq00002izr80-att/indonesia_contribution.pdf. Accessed 13 October 2020.

Karns, M. A., Mingst, K. A., & Stiles, K. W. (2004). The challenges of global governance. In M. P. Karns, K, A. Mingst, & K. W. Stiles (Eds.), *International organizations: The politics and processes* (pp. 1–42). Lynne Rienner Publishers, Inc

Kohama, H., & Teranishi, J. (1992) Nihon no ODA seisaku to hienjokoku no keizaihaten [In Japanese: Japan's ODA Policy and Economic Development of Recipient Countries]. In I. Yamazawa & A. Hirata (Eds.), *nihon amerika yoroppano kaihatsukyoryoku seisaku* [In Japanese: Japan / The USA / Europe's Development Cooperation Policy] (pp. 106–114). Institute of Developing Economies.

Mawdsley, E. (2012). *From recipients to donors: Emerging powers and the changing development landscape*. Zed Books.

MITI. (1987). *Keizaikyouryoku no genzyo to mondaiten* [In Japanese: Current status and problems of economic cooperation]. Tsusyo sangyo chosakai.

MOFA. (1981). *Keizaikyouryoku no rinen-seihukaihatsu enjo wa naniyue okonaunoka* [In Japanese: The Ideas of Economic Cooperation—Why has Official Development Assistance been conducted?]. MOFA.

MOFA. (1992). *ODA charter 1992*. https://www.mofa.go.jp/policy/oda/summary/1999/ref1.html. Accessed 11 November 2020.

MOFA. (2004). *Japan's Official Development Assistance White Paper 2004*. https://www.mofa.go.jp/mofaj/gaiko/oda/shiryo/hakusyo/04_hakusho/ODA2004/html/honpen/hp102020000.htm. Accessed 9 September 2020 and 2 February 2021.

MOFA. (2005). *Japan's Official Development Assistance White Paper 2005*. https://www.mofa.go.jp/policy/oda/white/2005/ODA2005/html/honpen/hp102020000.htm. Accessed 11 November 2020.

MOFA. (2006). *Betonamu ni okeru kaihatsu to wagakuni no taibetonamu enjo no doko* [In Japanese: Trends in development in Vietnam and Japan's assistance to Vietnam]. https://www.jica.go.jp/publication/monthly/0606/02.html#a03 Accessed 11 November 2020.

MOFA. (2010). *Japan's Official Development Assistance White Paper 2010*. https://www.mofa.go.jp/policy/oda/white/2010/pdfs/10_set.pdf. Accessed 11 November 2020.

Nishigaki, A., Simomura, Y., & Tsuji, K. (2009).*kaihatsuenjo no keizaigaku kyosei no sekai to nihon no ODA* [In Japanse: Economics of Development Assistance "The World of Symbiosis" and Japan's ODA]. Yuhikaku.

Nobayashi, K., & Naya, M. (Eds.). (2020). *Kikigaki Ogata Sadako kaiko-roku* [In Japanese: Memoirs' of Ogata Sadako]. Iwanamishoten.

Obuchi, K. (1998, December 16). *Policy speech by Prime Minister Keizo Obuchi at the Lecture Program hosted by the Institute for International Relations*. Toward the Creation of A Bright Future for Asia. https://www.mofa.go.jp/region/asia-paci/asean/pmv9812/policyspeech.html. Accessed 3 December 2020.

OECD. (1996a). *Shaping the 21st-century—The Contribution of Development Co-operation*. OECD.

OECD. (1996b). *OECD DAC List of ODA Recipients*. https://www.oecd.org/development/financing-sustainable-development/development-financestandards/historyofdaclistsofaidrecipientcountries.htm. Accessed 5 February 2021.

OECD. (2000). *OECD DAC List of ODA Recipients*. https://www.oecd.org/development/financing-sustainable-development/development-financestandards/historyofdaclistsofaidrecipientcountries.htm. Accessed 5 February 2021.

OECD. (2001). *OECD DAC List of ODA Recipients*. https://www.oecd.org/development/financing-sustainable-development/development-finance-standards/historyofdaclistsofaidrecipientcountries.htm. Accessed 5 February 2021.

OECD. (2003). *OECD DAC List of ODA Recipients*. https://www.oecd.org/development/financing-sustainable-development/development-finance-standards/historyofdaclistsofaidrecipientcountries.htm. Accessed 5 February 2021.

OECD. (2005). *Paris declaration on aid effectiveness*. OECD.

OECD. (2006a). *Applying strategic environmental assessment*. OECD.

OECD. (2006b). *OECD DAC List of ODA Recipients*. https://www.oecd.org/development/financing-sustainable-development/development-finance-standards/historyofdaclistsofaidrecipientcountries.htm. Accessed 5 February 2021.

OECD. (2007). *OECD DAC List of ODA Recipients*. https://www.oecd.org/development/financing-sustainable-development/development-finance-standards/historyofdaclistsofaidrecipientcountries.htm. Accessed 5 February 2021.

OECD. (2008). *OECD DAC List of ODA Recipients*. https://www.oecd.org/development/financing-sustainable-development/development-finance-standards/historyofdaclistsofaidrecipientcountries.htm. Accessed 5 February 2021.

OECD. (2010a). *OECD DAC List of ODA Recipients*. https://www.oecd.org/development/financing-sustainable-development/development-finance-standards/historyofdaclistsofaidrecipientcountries.htm. Accessed 5 February 2021.

OECD. (2010b). *Quality standards for development evaluation*. OECD.

OECD. (2012). *Busan partnership for effective development co-operation*. OECD.

OECD. (2014). *Accountability and democratic governance*. OECD.

OECD. (2021). *What is ODA?* https://www.oecd.org/dac/financing-sustainable-development/development-finance-standards/What-is-ODA.pdf. Accessed 1 March 2021.

OECF. (1991). Issues related to the World Banks approach to structural adjustment: a proposal from a major partner. *Kikin chosa kiho, 73*(1992.2), 11–18.

Short, C. (2004). *An honourable deception? New Labour, Iraq, and the Misuse of Power*. Simon & Schuster.
Suzuki, H. (2018). Kokusaikaihatsu kinyukikan no tsuzita nihon no chiken no hassin ni Kakawaru kosatsu: Sekaiginko no jirekenkyu ni motozuku zitaibunseki to tean [In Japanese: Consideration on the dissemination of Japanese knowledge through multilateral development banks: Fact-finding analysis and recommendations based on case studies of the World Bank]. *nihonno kokusaikyouryokuno rekisi* (1), 1–75.
Takahashi, M. (1998). Nihon no taikohatsukokumuke enjo no saikento: enjo no rinen to jijodoryokusien [In Japanese: Reexamination of Japan's Aid for least developed countries: Aid philosophy and self-help effort support]. In H. Imaoka (Ed.), *enjo no hyokato kokateki jisi, asiakeizaikenkyujyo* [In Japanese: Evaluation and effective implementation of aid] (pp. 73–119). Institute of Developing Economies.
Takahashi, M. (2015). Afurikakaihatsu enjo ni okeru nihon no yakuwari: Igirisuto no hikaku o tsujite [In Japanese: Japan's Role in African development assistance-through comparison with the United Kingdom Hikakuo tsujite]. In T. Kurozaki & K. Ozuka (Eds.), *Korekara no nihon no kaihatukyouryoku-bikudona kara sumato dona e, nihonhyoronsya* [In Japanese: Japanese international cooperation since now on-from big donors to smart donors] (pp. 65–95). Nihon Hyoronsha.
Takahashi, M. (2018). Nihon no enjo no shinkato taika: Kokusai kokyozen eno koke no kitei surumono [In Japanese: Evolution and Degeneration of Japanese aid: What is prescribed to contribute to the international public good]. *Keizai Shirin, 85*(4), 197–246.
Watanabe, T. (1991). Jijo doryoku shien no rinen no takaku kakageyo: Ko naki oda hihan no haisu [In Japanese: Raise the philosophy of self-help effort support high eliminate the criticism of ODA without reasons]. *Chuokoron, 106*(10), 85–96.
World Bank. (1993). *The East Asian Miracle: Economic growth and public policy*. Oxford University Press.

Interviews

The authors' interviews with aid officials from 2015 to 2021.
The interview with an anonymous Japanese scholar who has field experiences in international organizations (conducted on 18 April, 2021).

CHAPTER 6

Balancing Universal Values and Economic Interests Through Development Cooperation in Korea

Eunju Kim

1 Introduction

One of the remarkable features of official development assistance (hereafter, ODA) in South Korea is its high proportion of the economic development sector. When looking at support by sector, about 40% of total ODA in South Korea has been provided for economic infrastructure and services as of 2018, which is more than twice as high as the Organisation for Economic Co-operation and Development (OECD) Development Assistance Committee (DAC) average of 17% (OECD Stats). In addition, about one-fifth of bilateral aid is being used to support infrastructure (Donor Tracker).[1] Moreover, the proportion of concessional loans among

[1] https://donortracker.org/country/south-korea (last access November 1, 2020).

E. Kim (✉)
Hansung University, Seoul, South Korea
e-mail: eunjukim@hansung.ac.kr

H.-j. Kwon et al. (eds.), *International Development Cooperation of Japan and South Korea*,
https://doi.org/10.1007/978-981-16-4601-0_6

the total amount of ODA is 41.5% in South Korea, which is higher than that of other OECD/DAC member countries (OECD Stats). As of 2017, about 50% of loans to developing countries has come from agencies and companies in Korea, the highest among OECD/DAC member countries (OECD Stats; OECD, 2018). In this regard, the characteristics of ODA policy in South Korea, which has focused on economic development using a high proportion of concessional loans, have been criticized by civil society and academia (People's Solidarity for Participatory Democracy webpage).[2] The main reason for this criticism is the high proportion of tied aid and procurement primarily available to companies from the donor country (Eom et al., 2014; OECD, 2018). The benefits of foreign aid may therefore fall into the economic sector of the donor country and not into partner countries.

However, previous studies have shown that the characteristics of foreign aid are not just limited to the case of Korea. Literature comparing aid from Japan, China, and Korea has identified commonality among these three countries as "Asian donors" (Shimomura & Ohashi, 2013). It is characterized by providing aid mainly to the economic sector, with preferences for loans and an emphasis on mutual benefits or a so-called win–win situation (Reilly, 2012; Stallings & Kim, 2016). While some studies have revealed the characteristics of Asian donors, there are not many studies that have examined in-depth why each country has such characteristics. As such, it is necessary to go one step further by examining how they came to have such characteristics. Also, if these structural and fundamental features cannot be easily changed, efforts are needed to find a more desirable direction within the constraints.

Against this background, this chapter attempts to analyze economic-development-oriented ODA in South Korea. Why has foreign aid from South Korea been mainly focused on economic development? Does Korea's economic development ODA put more emphasis on the economic interests of Korea or the economic development of the partner country? What do we need to consider in terms of the future direction of economic development ODA while pursuing universal values such as economic development, poverty reduction, and quality of life in the partner country? In order to find the answers to the above questions, this chapter will first investigate the historical background of South Korea

[2] http://www.peoplepower21.org/PSPD_press/594794 (last access September 10, 2020).

as a recipient country, which affects the path of economic-development-oriented ODA policy. It will also explore policy changes in ODA from the Korean government officially established donor agencies in the late 1990s.

Second, it will examine whether Korea's economic-development-oriented ODA has pursued universal values or special interests. For this, it uses a theoretical framework that analyzes whether foreign aid policy is focused on universalism or specialism based on "accountability" (Kim, 2010). In the case of universalism, the scope of accountability is wide, while it is limited in the case of specialism. For example, universalism takes responsibility not only for the people of the donor country, but also the people of the recipient country and global society. Specialism takes responsibility by focusing on selected targets for support, especially for the selected countries and beneficiaries of individual projects. It furthermore gives more attention to domestic voters in the donor country. Kim (2010) defined the external conditions under which universalism and specialism appear. When the domestic political situation is stable and there is economic affluence at the same time, foreign aid policy close to universalism can be created. On the other hand, when domestic political or economic conditions are not stable, the tendency of specialism to emphasize the specific interests of the donor country increases. Using this theoretical discussion, this chapter will examine the scope of accountability of Korea's economic-development-oriented ODA in terms of whether it focuses more on accountability between the people in the recipient country or domestic voters. It also considers why such a trend has emerged considering the political and economic conditions in South Korea.

Third, based on the results of the analysis, it will suggest a way to balance universal values and special interests in Korea's economic-development-oriented ODA. It will consider how we can ultimately achieve national development in partner countries and contribute to supplying global public goods.

2 Economic-Development-Oriented Aid policy: Historical Path and Recent Policy Changes

The characteristics of Korean aid, which invests heavily in economic development, can be clearly seen when comparing total aid for economic development with total aid for social development. When Korea's foreign aid began in 1987, about 94% of ODA was provided for economic development. After the Korean International Cooperation Agency (KOICA) was established as an affiliate of the Ministry of Foreign Affairs in 1991, the number of grants increased, as did financial resources for social development. However, until 1995, 60% of total aid was still devoted to economic development. As shown in Table 1, aid for social development was 60% and aid for economic development was 21% in 2005, reversing the previous ratio. However, since 2010, aid for economic development has been steadily increasing, while the amount of aid for social development has decreased. In 2018, 40% of total aid was provided for economic development and 30% for social development, marking another point of divergence. As shown in Fig. 1, when including the production sector (agriculture, forestry, and fishing; mineral resources and mining; trade policies and regulations), 55% of aid was invested in economic development and production and 30% was invested in social development in 2018 (OECD Stats).

In order to examine why Korea's aid came to be centered on economic development, it is necessary to look at the historical background as a recipient country and the beginnings of official development assistance as a donor country. This can be traced back to the end of the Korean War. Immediately after the Korean War, Korea began to receive urgent

Table 1 South Korean aid provided to economic infrastructure and services (% of total ODA)

Sector	*1987 (%)*	*1990 (%)*	*1995 (%)*	*2000 (%)*	*2005 (%)*	*2010 (%)*	*2015 (%)*	*2018 (%)*
Social infrastructure and services	5.0	5.7	24.2	42.1	60.5	50.1	48.4	30.2
Economic infrastructure and services	93.8	91.3	60.6	47.7	21.0	33.7	32.9	40.2
Production sector			4.4	2.1	7.6	5.5	6.1	14.3
Multi-sector/cross-cutting				4.3	0.6	6.5	4.7	4.7
Commodity aid/general program assistance			3.2				0.1	0.1
Debt relief					0.6	0.1		
Humanitarian aid			0.2	0.2	5.6	1.1	1.9	4.2
Unallocated/unspecified	1.2	3.0	7.3	3.7	4.0	2.9	5.9	6.3
Total	100.0	100.0	100.0	100.0	100.0	100.0	100.0	100.0

Source OECD Stats (last access October 21, 2020)

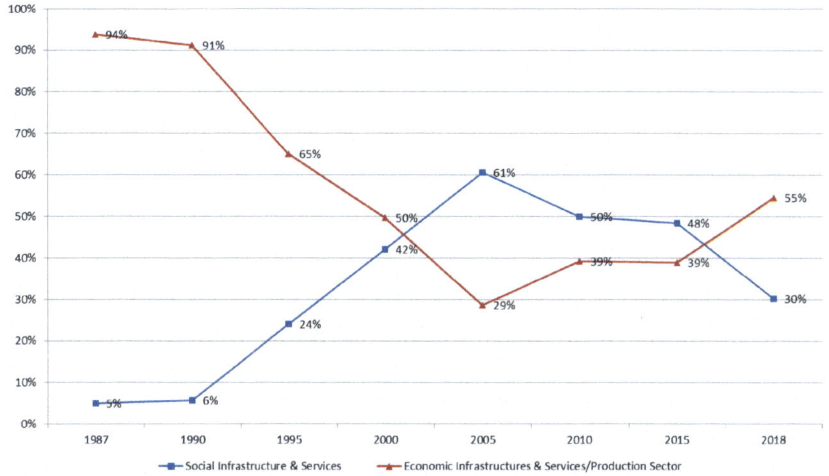

Fig. 1 South Korean aid provided to economic infrastructure and services/production sector (% of total ODA) (*Source* OECD Stats [last access October 21, 2020])

humanitarian assistance from the United States, specifically from United States Army Military Government (USAMGIK), through the Government Appropriations for Relief in Occupied Areas (GARIO). US aid began in the late 1950s, and for 10 years from 1953 to 1962, the US provided a total of USD 2 billion in economic aid and USD 1 billion in military aid. As a result, the impact of US aid on the Korean economy became significant, with US aid making up 70% of Korea's total income and 75% of total fixed capital between 1953 and 1961. On the other hand, Japanese aid to Korea began from the 1960s. With the resumption of diplomatic relations between South Korea and Japan in the 1965, the South Korean government received JPY 300 million in grants, JPY 200 million in loans, and JPY 300 million in private finance based on the Korea–Japan Agreement. Also, starting with the railway facility reconstruction project in 1966, yen-denominated loans began, and Japanese aid continued, focusing on infrastructure construction. A total of 91 projects, including the construction of the Seoul subway, dams, roads, water and sewage pipelines, educational facilities, and medical institutions, as well as infrastructure construction projects worth JPY 500 billion, were carried out through the Japan Overseas Economic Cooperation Fund

(OECF) (Kondoh, 2013). It is true that until the 1960s and 1970s, which marked a period of rapid economic growth in Korea, financial resources through foreign aid became a major driver of Korean economic development. There is no denying that foreign aid had a significant impact on the Korean economy until the World Bank's loans were repaid in 1995, ending Korea's history as a recipient country.

In this way, through foreign aid, economic development in Korea was achieved at a rapid pace. Furthermore, South Korea went through a successful transition from a recipient to a donor country. In this line, Korea's history as a recipient country after the Korean War would have a significant impact on Korea's policy decisions after being a donor country. It can be argued that this historical path has also influenced Korea's tendency to invest in economic development after becoming a donor country. The same argument could be made for Korea's decision to provide funds to developing countries in the form of concessional loans as it contributed significant financial resources toward the economic development of South Korea.

Secondly, in the 1980s, while promoting export-led economic growth, Korea enjoyed a so-called third low boom with low interest rates, low oil prices, and a low value of the won, thereby securing a considerable amount of foreign exchange reserves. As Korea reached a current account surplus from 1986, Korea's economic influence in the region began to grow (Chun, 2018). At the time, the administration was looking for a way to use foreign exchange reserves and discovered the Japan Bank for International Cooperation (JBIC) model of Japan's export–import bank. The Korean government learned that there was a way to provide financial support to developing countries in need of resources for economic development, while at the same time fulfilling the demands of Korean companies that wanted to export and develop overseas markets abroad. Japan's Overseas Economic Cooperation Fund (OECF) was set as a benchmarking model, and the Economic Development Cooperation Fund (EDCF) was established at the Export–Import Bank after its name.[3] In 1987, when Korea created the Economic Development Cooperation Fund (EDCF) at the Export–Import bank, a subsidiary of the Ministry of Strategy and Finance, Korea's ODA first started out its donorship in the form of concessional loans. In this context, from the very beginning,

[3] Interview with an Export–Import Bank official (August 21, 2020).

Korea's ODA was meant to serve two purposes: supporting development resources necessary for economic development in developing countries and seeking a new path for the export-oriented Korean economy.

In summary, it can be said that there were two historical reasons for providing aid mainly in the sector of economic development before 2010. One is Korea's distinctive historical background as a recipient country, as Korea achieved economic development through foreign aid. The other is that Korea began in earnest as a donor in the late 1980s by providing aid in the form of concessional loans.

Then, why did the share of the economic development sector surge again after 2010? It is related to the policy changes that occurred after Korea joined the OECD/DAC in 2010, such as the systemization of ODA policy by sector and region.

First, sectoral ODA policies gradually value economic development as a priority goal. Since 2010, when Korea joined OECD/DAC, the ODA policy in Korea has set out policy goals by sector, region, and income group. After joining the OECD/DAC, the First Strategic Plan for International Development Cooperation (2010–2015) was launched. According to the bilateral aid strategy for each field, the priority for Korea's aid was set in the following order: green growth, economic infrastructure, and social infrastructure (Government of Republic of Korea, 2010: 7). Green growth, which was promoted by the Lee Myung-bak administration as a major domestic policy at the time, was applied to ODA as well. Moreover, it was announced that the administration would provide 30% of the EDCF support for green growth industries (solar power, wind power, bioindustry, water and sewage/wastewater treatment, waste treatment, small-scale hydropower). In addition, a plan was established to support the construction of economic infrastructure, for which the greatest demand came from developing countries, such as transportation and energy. It is true that the real intention of green growth mainly focuses on *growth* rather than environmental protection and climate change. Because it was actually close to infrastructure construction, environmental issues became secondary to economic growth in the discourse on green growth in Korea at that time. In other words, the aid ratio in the economic sector increased again after 2010 as policy goals prioritized economic growth and infrastructure construction.

In addition, the Second Strategic Plan for International Development Cooperation (2016–2020) placed more emphasis on the expansion of economic infrastructure. The Second Strategic Plan stated that the government would continue to support health, education, and rural development sectors related to social development, but would expand support for economic infrastructure and environmental sectors highlighted in the Sustainable Development Goals (SDGs) (Government of the Republic of Korea, 2016: 4). While discussing the post-2015 agenda, the limitations of the MDGs and criticisms regarding their narrow focus on social development were addressed, and targets encompassing economic development and the environment were also included in the SDGs. Such global discussions also had a significant impact on the ODA policy in Korea, which led to setting economic development as a priority for ODA policy. Thus, the sudden increase in the sector of economic development and production after 2015, as shown in Fig. 1, could be a result of such policy changes.

Second, strategic regional policy for ODA has moved the center of aid back to economic development because it focuses on the Asian region where demand for economic development has been on the rise. Since Korea established government agencies for implementing ODA in the late 1980s, the continent that has been supported the most is Asia. However, in the First and Second Strategic Plan for International Development Cooperation, it was decided to reduce support for Asia and gradually increase support for Africa. As a result, the proportion of support to the Asian region gradually decreased from 62% in 2011 to 47% in 2018. The proportion of support to Africa was 18% in 2011 but gradually increased to 27% in 2018 (OECD Stats). However, more than half of Korea's ODA is still concentrated in Asia as the strategic priority region of Korea's ODA policy. Moreover, in 2017, the Moon Jae-in administration announced the New Southern Policy as a major foreign policy and promised to strengthen exchanges and cooperation with Association of Southeast Asian Nations (ASEAN) countries and double ODA support. After that, aid to Asia increased again, and in 2019, the share of aid to Asia rose to 51% and aid to Africa fell slightly to 25% (OECD Stats).

Most of the countries in Asia are now in the process of leaping from underdeveloped countries to middle-income countries. For example, countries such as Vietnam, Indonesia, and the Philippines have graduated from least-developed country (LDC) status and entered the status of lower middle-income countries (LMICs). The most important point in the national development strategy of Southeast Asian countries in terms of becoming middle-income countries is economic development. Thus, the development needs of partner countries in Asia are mostly related to infrastructure building for transforming their economies. In addition, in order to prevent themselves from falling into the middle-income country trap, whereby they stay low- and middle-income countries, there is a sizable demand for infrastructure construction. These include energy, transportation, agricultural irrigation systems, natural disaster prevention infrastructure, and water quality and sanitation infrastructure (World Bank, 2019). Moreover, as the SDGs include economic development goals, many developing countries consider economic development a top priority and require support in economic development and infrastructure building. In addition, the "One Belt, One Road" initiative of the Chinese government is stimulating demand for infrastructure such as roads, railroads, and port development with an ambitious goal of connecting to Africa beyond Asia. In response, Japan is also expanding large-scale support by promising "quality infra" to Southeast Asian countries. Korea also finds itself positioned between the large-scale economic development and infrastructure support competition between China and Japan. This may also result in the allocation of more financial resources to economic development in order to catch up in terms of aid competition in Asian regions. Considering the demands of these Asian countries and aid competition in the region, it seems that ODA policy for the Asian region has mostly been centered on areas related to economic development.

With Korea's emphasis on the economic sector and a strong tendency to focus on Asia in the process of systematizing ODA policy after joining the OECD/DAC in 2010, foreign aid focusing on the economic sector has increased again.

3 Was It for Universal Values or Special Economic interests?

It has been argued that Korea is giving aid mainly to the economic development sector. If this is indeed true, what we need to reconsider is the question of whether it is for universal values or for the special interests of the donor country. As discussed in the Introduction, we can distinguish between *universalism* and *specialism* in international development based on the scope of accountability. In the case of specialism, there is a tendency to place more emphasis on domestic voters or domestic political constituencies instead of widening the object of policy accountability toward the people in partner countries or in global society (Kim, 2010).

In this respect, this section will reflect on whether Korea's economic-development-oriented ODA takes responsibility for the people in partner countries or the constituencies in the donor country.

Amid the prolonged economic recession since the 2008 global financial crisis, so-called aid nationalism is gradually increasing across the world (Kharas & Rogerson, 2017). This means that it is important not only for the recipient country but also for the donor country to receive benefits through aid. For the donor country, the pursuit of economic and security interests of its own has increased in particular. Against this backdrop, traditional donors such as the US, the UK, Australia, European countries, and Japan are putting their national interests forward in order to justify their aid budgets considering domestic constituencies (Gulrajani, 2017; Gulrajani & Swiss, 2017).

South Korea's economic-development-oriented ODA policy is also on the way to leaning toward the economic interests of the donor country. In the policy documents of the South Korean government, the economic interests of South Korea are put ahead of reducing the gap between countries in global society. In 2017, the Moon Jae-in administration announced the Presidential Policy Agenda, which included the agenda for ODA. The main goal would be "strengthening economic diplomacy and development cooperation to enhance national interests" (Government of the Republic of Korea, 2017). Following the Moon Jae-in administration's emphasis on national interests in foreign affairs, each government department involved in ODA has repeatedly highlighted national interests in their own ODA policies and projects. For example, in 2020, in the Annual Implementation Plan for International Development Cooperation, the main goal of ODA projects was set as "promoting ODA that contributes to the national interest" by the Ministry of Foreign Affairs. Among other governmental ministries related to economic development, the Ministry of Strategy and Finance has set "support for the economic development of partner countries and strengthening economic cooperation for win–win growth" as its main aid objectives. The Ministry of Commerce, Industry, and Energy will pursue "strategic cooperation in the industry and energy sector with developing countries", while the Ministry of Agriculture, Forestry, and Livestock aims to create "positive relations to pioneer new overseas markets for the ministry". Among the ministries related to social development, the Ministry of Education has stated that it will "enter the international platform through educational ODA with domestic experts, universities, and research institutes". The Ministry of

Health and Welfare, on the other hand, will "reinforce contributions to the international community while promoting national interests". The Cultural Heritage Administration has also set "job creation through cultural ODA and the promotion of national interests" as the main goal of its ODA projects (Government of the Republic of Korea, 2020). In other words, among government departments and public institutions participating in Korea's ODA, the tendency to place more importance on economic interests while considering domestic constituencies is growing.

The reason for this trend is due to changes in external conditions related to the political and economic situation. Kim (2010) insists that the trend of universalism will appear when the ODA budget increases amid economic prosperity. On the other hand, specialism appears when political and economic conditions are not stable (Kim, 2010). In this line, specialism focusing on domestic economic interests has appeared in Korea, according to which it has expanded the aid budget amid economic recession.

The total ODA budget in Korea has been increasing at a rapid pace since joining OECD/DAC in 2010. Furthermore, it is expected to increase steadily until 2030 when the policy goal of 0.3% of the gross national income (GNI) to ODA promised to the international community is achieved. However, as the Korean government faces more domestic economic challenges, there is intense debate among government ministries on whether or not it should increase the ODA budget. In particular, if the COVID-19 pandemic leads to a recession in the global economy, it is unlikely that the ODA budget will increase steadily due to a deficit in the government budget. Recently, during discussions related to the Third Strategic Plan for International Development Cooperation, the Ministry of Strategy and Finance insisted on readjusting the target to increase the ODA budget to 0.3% of the GNI by 2030 based on the economic outlook and mid-term fiscal management plan. In response to this opposition, the Ministry of Foreign Affairs and other related government ministries have emphasized the ripple effect of ODA on the domestic economy in order to persuade the Ministry of Strategy and Finance as well as the National Assembly.[4]

[4] The author was able to engage in the policy process for drafting the third strategic plan. This observation is based on the author's experiences as a member of the task force in 2020.

Moreover, as unemployment has become a serious issue in domestic policy, the Moon Jae-in administration has stated that "increased development assistance that contributes to employment and national interests" will be placed on presidential policy agendas. Specifically, policy tasks included "contributing to the creation of employment in the private sector through collaboration with companies and cultivation of global talent, creation of jobs in the public sector in the ODA field, and contribution to national interests through overseas expansion such as infrastructure projects" (Government of the Republic of Korea, 2017).

In addition, in order to persuade domestic constituencies and representatives of the National Assembly, public officials continue to emphasize the fact that aid has domestic returns to the Korean economy. One of the common arguments presented to persuade government ministries and domestic voters is that, through aid, Korean companies can subsequently enter developing countries, which can lead to an increase in exports.[5] Korea has traditionally pursued export-led economic growth and has always turned its eyes to overseas markets. As economic growth in Korea continues to slow down, this policy direction is becoming ever more explicit. The government's ODA support plans and policies also emphasize the fact that Korean companies can enter overseas markets and increase exports through aid.

Against this background, it can be said that the tendency toward specialism in South Korea has grown by taking narrow accountability focused on selected targets of individual projects and domestic voters in the donor country. As a result, according to the principal aid index recently released by the UK Overseas Development Institute, South Korea scored lower than in other dimensions by pursuing short-sighted domestic economic returns through aid.[6] The fact that ODA is increasingly being recognized for the purpose of boosting Korea's economic interests and creating jobs is an area of future concern. Development assistance focused on the special interests of donors may be far from the original purpose of international development cooperation to reduce the gap between the rich and the poor through economic growth and poverty eradication (Escobar, 1995). International development cooperation should not

[5] Interview with government officials (November 15, 2020).

[6] Overseas Development Institute, Principal Aid Index 2020, https://www.odi.org/opinion/10502-principled-aid-index (last access November 9, 2020).

narrowly pursue the national interests of a single country but aim to protect global public goods and contribute to the common interests of other countries.

If the donor country and the recipient country first seek mutual benefits while pursuing global public goods, the ripple effect would be helpful to Korea's economic and security interests. It is necessary to clearly distinguish between the transposition of purpose and means as well as to acknowledge the order of direct and ripple effects. In other words, the underlying principles of ODA policy for economic development should be directed at supplying global public goods while balancing the economic interests of donor and partner countries.

4 The Way Toward Private Sector Engagement for Mutual Prosperity

We need to pursue economic development aid that can create mutual prosperity for donor and partner countries. To do so, first and foremost, economic development aid policy needs to pay attention to private sector development in developing countries. Considering the feeble size of the private market, attention should be paid to revitalizing private markets and actors in developing countries, developing private enterprises with appropriate business activities, and eventually creating jobs for the people in developing countries. In this respect, ODA policy needs a focus on supporting financial and business activities in developing countries through aid. However, Korea's economic development aid since 2010 has shown increased support for transportation and energy infrastructure. This is because Korea's economic development aid has overly emphasized the advancement of Korean companies abroad, increasing exports for Korea, and job creation for Koreans.

To support enlarging the private market, banking and financial services, as well as business and other services, account for very little in the total aid of South Korea, even within the economic infrastructure and services sector. In 2018, financial services and business support combined made up USD 10 million, only accounting for 0.5 and 0.2% of ODA, respectively (as shown in Fig. 2). Compared with European countries that also provide concessional loans, financial services and corporate support accounted for 6.1% (USD 1,373 million) in Germany and 4.6% (USD 437 million) in France, marking a sizeable difference from Korea (Donor Tracker). Therefore, it is necessary to diversify economic development aid

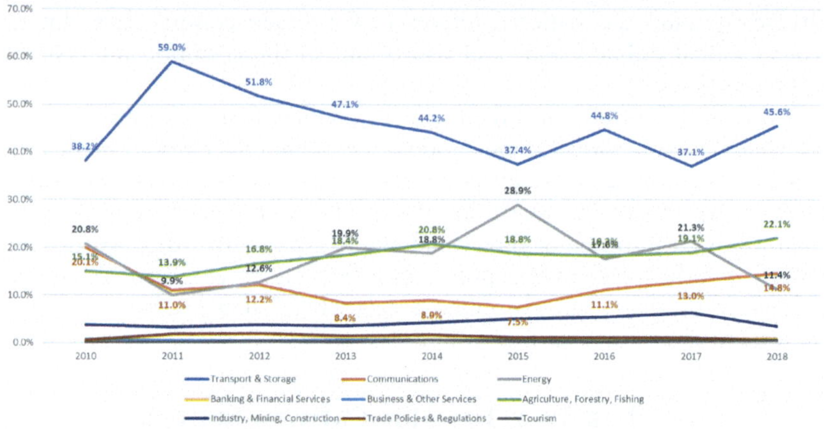

Fig. 2 Economic infrastructure and production sector in South Korea by component (2010–2018) (*Source* OECD Stats [last access October 21, 2020])

policies of Korea in the direction of revitalizing various production sectors by breaking away from the infrastructure-oriented trend.

Secondly, economic development aid policy needs to be shifted in the direction of expanding the private sector engagement of the donor and partner country. *Private sector engagement through development* refers to participation in international development cooperation with the aim of contributing to sustainable development by addressing economic and social issues in developing countries (Di Bella et al., 2013). Efforts should be directed at building an inclusive business model and a sustainable value chain with partner countries. As shown in Fig. 3, the concepts related to the private sector and development cooperation can be distinguished and divided into three stages. South Korea is currently at a stage in which private sectors are participating in development cooperation through general commercial activities. This is the so-called "private sector in development" stage. However, it is necessary to shift to the stage of "private sector engagement" by contributing to sustainable development, which goes beyond general commercial and profit-seeking activities.

Thirdly, more highlight should be focused on private companies in developing countries. As mentioned before, Korea's economic development and production sector aid are focused on the construction of transportation and energy infrastructure, and the main participants in

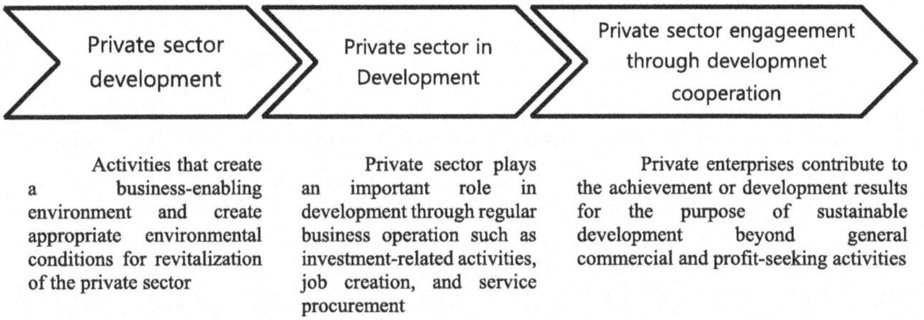

Fig. 3 Definition of private sector engagement through development cooperation (*Source* author's interpretation based on Di Bella et al. [2013])

such projects are primarily engineering companies in Korea that attain procurement by benefiting from tied aid. In fact, the number of companies that undertake infrastructure projects after successfully completing various aid projects in developing countries is increasing. Economic development aid projects play the role of a catalyst for domestic companies to enter markets in developing countries. The EDCF of the Export–Import Bank, which operates Korea's concessional loans, views this as a great achievement by boosting the number of domestic companies in global ODA markets (EDCF, 2020). However, the means and purposes of aid should never be confused. In other words, while Korean companies may participate in aid projects, the main focus should not be on creating profits for Korean companies or new opportunities to enter overseas markets. The priority should be to ensure that participating enterprises contribute positively to the development of private enterprises and private markets in developing countries. If private companies are ever involved in an aid project, it should support the growth of private companies in developing countries and ensure that they can play a critical role in the global supply value chain. Thus, more emphasis should be placed on supporting private enterprises in developing countries rather than domestic companies in the donor country.

In order to implement these principles in the field, policy changes are needed in strategies and norms, governance, and policy tools in economic development aid policy (Kim & Lee, 2019). First of all, strategies and norms for participation in the private sector should be established and shared with various stakeholders in the field of development cooperation. South Korea should also pursue goals related to economic development in developing countries through the expansion of private sector engagement and share this vision among various stakeholders. South Korea should move away from the stage of "private sector in development" that views the partner country as the object and the donor country's enterprises as the subject of aid. Future strategies should instead encourage the engagement of private enterprises in the donor and partner countries to create and develop common interests.

In terms of governance, the South Korean government should build a platform to accelerate collaboration between governments and companies. In order to enhance private sector cooperation and financial support, the voices of various actors hidden in the development cooperation community should be heard through a business partnership platform. If Korea was to create such a business partnership platform, the role of the government should be a coordinator and facilitator to ensure that partnerships with private companies operate well. In the early stages, the government should present enterprises with the main problems faced by developing countries and provide the necessary information for business activities in developing countries. In addition, a matching fund should be provided for projects with high development effectiveness so that ODA resources could be used as a catalyst. Enterprises, on the other hand, would use their technology to solve various problems in developing countries and mobilize more private resources to contribute to the development of the private sector in developing countries. If a company was to initiate business activities and create a leading company that performs successfully, this network could further enable information sharing and cooperation between companies.

In addition, it is possible to change policy tools to expand financial support and knowledge sharing. It is necessary to diversify the policy tools of financial support by establishing development finance institutions (DFIs) and providing stock investment, guarantees, and risk insurance. In traditional donor countries, existing development agencies are now readjusting their functions and roles or establishing new development financial institutions to meet gaps in financial resources. Most OECD/DAC member countries have established a separate development finance institutions to mobilize necessary resources for developing countries and encourage private companies to participate in development cooperation. On the other hand, South Korea is still at the stage of ideation. It needs to begin discussions on establishing a development financial institution with which to utilize development funds as leverage for increasing financial resources. This would also enable Korea to mobilize more private resources through the diversification of financial support means.

Moreover, information sharing can be a critical policy tool to induce the transfer of private innovative technologies and contribute to the development of developing countries. Although companies with innovative technology are offering solutions to diverse problems in development cooperation, information on local problems that occur within the developing country is still insufficient. In this case, donor agencies can function as an intermediate link for private companies by providing necessary knowledge.

5 Conclusion

Korea's foreign aid has been characterized as focusing on economic development, providing loan packages, and emphasizing the mutual benefits of the donor–recipient relationship. Since these characteristics can be found not only in Korea, but also in Japan and China, some scholars have identified them as the characteristics of Asian donors. Previous studies have examined Korea's aid behavior as an Asian donor using evidence before 2010, but the trend has grown even greater since 2010 after joining the OECD/DAC. Therefore, this study examines historical institutions and recent policy changes after 2010 at the same time in order to explain the reason that Korea has such characteristics. This study also analyzed whether South Korea takes accountability for the global society or domestic voters in its economic development aid. It was determined that the tendency toward specialism, which focuses on the beneficiaries of specific projects and/or emphasizes domestic voters, has been growing in the recent context of aid nationalism among traditional and emerging donors.

South Korea needs to put more efforts into balancing universal and special interests while continuing to pursue economic-development-oriented ODA. Unlike the Millennium Development Goals (MDGs), which focused on social development, the international community has included new goals for economic development in the 2030 SDGs. For example, Goal 8 of the SDGs includes "decent work and economic growth", while "industry, innovation and infrastructure" are presented in Goal 9. As such, the aid architecture and various actors for international development cooperation are rapidly changing, but South Korea is still focusing only on ODA led by the public sector. Further policy efforts should focus on establishing a comprehensive aid strategy that also includes the distinctive role of the private sector. International development cooperation in Korea needs to develop into an "aid ecosystem" that encompasses not only public agents but also private agents. More importantly, partner countries must be considered equally and simultaneously within the ecosystem so that the donor and partner countries can contribute to mutual prosperity.

In this line, the economic interests of the donor country should be redefined in Korea's international development cooperation. As De Tocquville suggested, individuals' short-term interests and long-term interests should be classified in the direction of pursuing public interests in society (De Tocqueville, 2003; as cited in Gulrajani & Calleja, 2019). This could also be applied to economic interests pursued in international development cooperation. Even if individual countries pursue their own national interests through aid, they must also maintain an interest in and philosophy toward global public goods. Policy decisions should be based on the principle of pursuing the interests of individual countries in line with the global public good. The economic benefits of the donor country that may or may not arise in the process must be separately classified as incidental ripple effects. The ultimate goal of international development cooperation is the economic and social development of developing countries as well as a contribution to supplying global public goods.

References

Chun, B. Y. (2018). An exploratory study on the structural changes of growth factors of the Korean economy after 1987. *Citizen and the World, 31*, 93–121.

De Tocqueville, A. (2003). *Democracy in America* (Vol. 10). Regnery Publishing.

Di Bella, J., Grant, A., Kindornay, S., & Tissot, S. (2013). *Mapping private sector engagements in development cooperation*. North–South Institute. http://www.nsi-ins.ca/publications/mapping-private-sector-engagements-in-development-cooperation/ (last access 10 April 2021)

EDCF. (2020). *Performance of EDCF in 2019*. Internal document of EDCF

Eom, Y., Ho, J. H., & Heon, J. J. (2014). An empirical analysis of the determinants of tied South Korean aid. *Korean Journal of Public Administration, 52*(1), 123–144.

Escobar, A. (1995). *Encountering development: The making and unmaking of the Third World*. Princeton University Press.

Government of the Republic of Korea. (2010). *The first international development cooperation strategic plan 2010–2015*. https://www.odakorea.go.kr/ODAPage_2018/category02/L02_S01_02.jsp (last access 10 April 2021)

Government of the Republic of Korea. (2016). *The second international development cooperation strategic plan 2016–2020*. https://www.odakorea.go.kr/ODAPage_2018/category02/L02_S01_02.jsp (last access 10 April 2021)

Government of the Republic of Korea. (2017). *Presidential policy agenda*. https://www.evaluation.go.kr/psec/np/np_2_1_2.jsp (last access 10 April 2021)

Government of the Republic of Korea. (2020). *Annual international development operation plan*. https://www.odakorea.go.kr/hz.blltn2.YearPlanSlPL2.do?brd_seq=3&blltn_div=oda (last access 10 April 2021).

Gulrajani, N. (2017). Bilateral donors and the age of the national interest: What prospects for challenge by development agencies? *World Development, 96*, 375–389.

Gulrajani, N., & Calleja, R. (2019). *Understanding donor motivations: Developing the principled aid index*. Overseas Development Institute.

Gulrajani, N., & Swiss, L. (2017). *Why do countries become donors? Assessing the drivers and implications of donor proliferation* (ODI Report). Overseas Development Institute.

Kharas, H., & Rogerson, A. (2017). *Global development trends and challenges: Horizon 2025 revisited*. Overseas Development Institute.

Kim, E. J., & Lee, D. S. (2019). *Job creation in international development by strengthening the ODA Ecosystem*. Korea Institute of Public Administration.

Kim, T. (2010). The dual structure of value orientations for international development cooperation: A case of Japan. *Review of International and Area Studies, 19*(2), 67–104.

Kondoh, H. (2013). Korea's pathway from recipient to donor: How does Japan matter? In J. Sato & Y. Shimomura (Eds.), *The rise of Asian donors: Japan's impact on the evolution of emerging donors*. Routledge.

OECD. (2018). OECD development co-operation peer reviews: Korea 2018. *OECD Development Co-Operation Peer Reviews, OECD Publishing*. https://doi.org/10.1787/9789264288829-en. (last access 10 April 2021)

Reilly, J. (2012). A northeast Asian model of ODA? Comparing Chinese, Japanese and Korean official development assistance. In C. Dent & J. Dosch (Eds.), *The Asia-Pacific, regionalism, and the global system* (pp. 216–235). Edward Elgar.

Shimomura, Y., & Ohashi, H. (2013). Why China's foreign aid matters? In Y. Shimomura & H. Ohashi (Eds.), *A study of China's foreign aid*. Palgrave Macmillan.

Stallings, B., & Kim, E. M. (2016). Japan, Korea, and China: Styles of ODA in East Asia. In H. Kato, J. Page & Y. Shimomura (Eds.), *Japan's development assistance*. Palgrave Macmillan.

World Bank. (2019). *Beyond the gap: How countries can afford the infrastructure they need while protecting the planet*. https://documents.worldbank.org/en/publication/documents-reports/documentdetail/189471550755819133/beyond-the-gap-how-countries-can-afford-the-infrastructure-they-need-while-protecting-the-planet (last access 10 April 2021)

Press and Online Resources

Donor Tracker. https://donortracker.org/country/south-korea (last access 20 October 2020)

OECD Stats. https://stats.oecd.org/ (last access 21 October 2020)

Overseas Development Institute, Principal Aid Index. https://www.odi.org/opinion/10502-principled-aid-index (last access 9 November 2020)

People's Solidarity for Participatory Democracy. http://www.peoplepower21.org/PSPD_press/594794 (last access 10 September 2020)

CHAPTER 7

New Partnership with the Private Sector in Japanese Development Cooperation

Izumi Ohno and Sayoko Uesu

1 Introduction

The landscape of international development in the age of the Sustainable Development Goals (SDGs) is significantly different from that in the time of the Millennium Development Goals (MDGs). The SDG age is characterized by more diverse actors, including the private sector. With accelerated globalization, the flow of private financing to developing countries has dramatically increased, and now exceeds the volume of official development assistance (ODA). In 2017, private finance accounted for roughly

I. Ohno (✉) · S. Uesu
National Graduate Institute for Policy Studies (GRIPS), Tokyo, Japan
e-mail: i-ohno@grips.ac.jp

S. Uesu
e-mail: S-Uesu@grips.ac.jp

H.-j. Kwon et al. (eds.), *International Development Cooperation of Japan and South Korea*,
https://doi.org/10.1007/978-981-16-4601-0_7

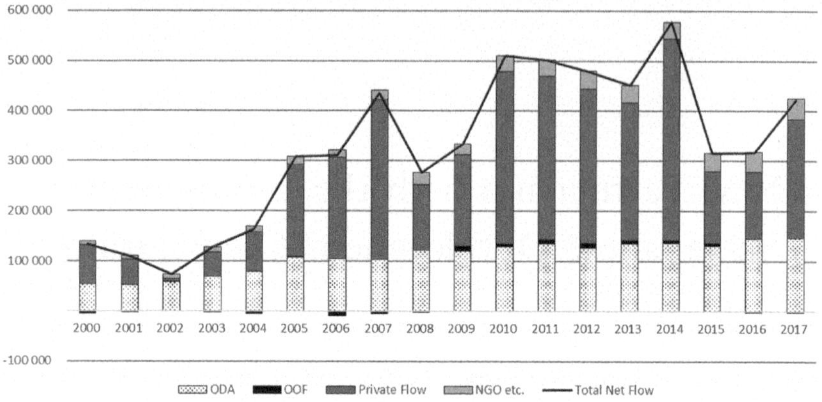

Fig. 1 Financial flows from OECD-DAC countries to developing countries (USD million) (*Source* author's compilation, based on OECD-DAC statistics)

60% of total financial flows[1] to developing countries from member countries of the Organization for Economic Co-operation and Development (OECD) Development Assistance Committee (DAC), while ODA constituted only one-third (Fig. 1). As a result, the interface between business and development cooperation activities is growing, and major donors have begun to introduce new instruments for collaboration with the private sector. Furthermore, with the expansion of global value chains (GVCs) in low-income countries, multinational corporations (MNCs) are increasingly urged to ensure decent work, social, and environmental compliance. A global coalition for sustainable production and manufacturing is emerging, involving businesses, donor agencies, and civil society organizations (CSOs). More recently, the COVID-19 pandemic which broke out in December 2019 presents a convincing argument for the need to take a human-centered approach. The role of the private sector has become even more vital for ensuring a post-pandemic recovery that "builds back better" and making economic, social, and environmental progress.

[1] Based on OECD-DAC statistics, ODA, non-ODA flows (other official flows (OOF), FDI, net grants by NGOs and foundations) are included.

From around 2010/11, the Japanese government and the Japan International Cooperation Agency (JICA) also introduced a number of new measures to promote partnerships with the private sector within its ODA activities, aiming at the utilization of Japanese private expertise and technologies for addressing development challenges in emerging economies and developing countries. This direction is clearly articulated in the Development Cooperation Charter adopted by the Cabinet in 2015. At the same time, it should be noted that for Japan, partnership between business and development is not a new phenomenon. For decades, Japan has provided East Asia with cooperation that combines trade, investment, and ODA (the so-called Japan ODA Model) (METI, 2005; Shimomura & Wang, 2013), contributing to the development of dynamic regional production networks. How do these new partnerships differ from the public–private partnerships (PPPs) of the past? What does this type of development hold for the future?

The rest of this chapter is divided into four parts. Section 2 reviews recent global trends and new models of development partnerships with the private sector. Section 3 examines the evolution of Japan's development partnership with the private sector, particularly the country's new ODA initiatives introduced over the past decade. Section 4 presents case studies of new ODA initiatives in partnership with the private sector. The final section considers implications of new development partnership models and the way forward for Japan, in light of the need for effective contributions to the post-pandemic recovery and to the achievement of the SDGs.

2 Recent Global Trends and New Models of Development Partnerships with the Private Sector

Since the early 2000s, the international development community has introduced a great number of new measures to promote partnerships with the private sector within ODA activities. Three types of development partnership are worthy of note: (i) Base of the Pyramid (BOP) business, alternatively called "inclusive business," which is a private sector approach to providing goods, services, and livelihoods to low-income groups on a commercially viable basis; (ii) the SDGs business, or "sustainable development business," which provides business solutions for broader global development challenges; and (iii) corporate social responsibility (CSR) in value chain management, which addresses a wide range of issues, with substantial focus on such aspects as decent work, safety, the environment, and procurement practices.

Behind such moves was the initiative of the private sector, which seized the opportunity presented by accelerated globalization. Especially after the 2008 financial crisis, MNCs actively expanded their business operations in low-and middle-income countries, shifting their investments away from stagnant advanced countries. In addition, the international business and financial communities faced growing calls to pay due attention to societal progress as a core business strategy, not just a means of short-term profit maximization (Porter & Kramer, 2011).[2] The crisis prompted a number of donor countries to seek potential synergies between development aid and domestic commercial interests (Groɞe-Puppendahl et al., 2016).

Furthermore, due to technological progress and the reduced costs of transport and telecommunication, production processes are becoming increasingly fragmented, arranged in GVCs or global networks spanning many countries (Barrientos et al., 2011). With the expansion of GVCs in low-income countries, MNCs have been strengthening an increasingly wide range of CSR activities, such as the promotion of decent work, and social and environmental compliance. Such trends correspond to the vision embraced by the 2030 Agenda for Sustainable Development or the SDGs, adopted at the United Nations (UN) in 2015. A global coalition for sustainable production and manufacturing is emerging, bringing together businesses, donor agencies, and CSOs to promote the setting of standards for CSR and also to monitor MNC supply chain management.

2.1 BOP Business

BOP business, or "inclusive business," is a new development partnership model in which core business operations are designed to directly address the social and development challenges faced by low-income consumers in developing and emerging economies (Prahalad, 2004). The concept of BOP business was originated in the late 1990s by Hart and Prahalad, who presented it as "strategies for the bottom of the pyramid" (Prahalad &

[2] Porter and Cramer argue that the principle of shared value involves creating economic value in a way that *also* creates value for society by addressing its needs and challenges. Shared value goes beyond social responsibility and philanthropy and is a new way to achieve economic success.

Hart, 1999). According to the International Financial Corporation (IFC), the BOP sector of the population comprises four billion people who earn less than USD eight per day in purchasing power parity (Hammond et al., 2007). Originally, corresponding to the international commitment to achieving the MDGs by 2015, BOP business came to attract attention as an innovative, market-based approach to poverty alleviation, avoiding aid dependency.

A classic example is Hindustan Unilever (a subsidiary of the British-Dutch corporation Unilever), which sells consumer products (e.g., soap, shampoo, water purifiers) in small packages at affordable prices to contribute to the improvement of health and sanitation in rural India. By engaging rural women as door-to-door sellers or managers of petty shops, Hindustan Unilever supports the empowerment of women. It also collaborates with donors including the United States Agency for International Development (USAID), World Bank, and the United Nations Children's Fund (UNICEF) to raise awareness of health and hygiene issues through programs such as Global Public-Private Partnership for Handwashing with Soap.

The BOP business has inspired numerous new projects supported by donors and the private sector, but it was criticized early on for being too commercially driven, given its weak commitment to upgrading local industry/economic activities and the promotion of skill transfer to the poor. The subsequent BOP 2.0 was aimed at addressing that problem by engaging the low-income group not just as consumers, but also as producers, distributors, and suppliers. The BOP 2.0 business model has also helped to strengthen supply chains for business (Ghosh & Rajan, 2019).

2.2 *Public-Private Partnerships for Sustainable Development*

The second new type of development partnership, broader business engagement aimed at sustainable development, reaches beyond poverty reduction. In a sense, this approach echoes the shift in global development from the MDGs to the SDGs. The 2008 financial crisis prompted global calls for responsible and sustainable business, and the adoption of the SDGs by world leaders at the UN in 2015—which in turn have greatly contributed to the promotion of PPPs for sustainable development.

The SDGs mark a departure from the MDGs in that they call on business explicitly to apply their creativity and innovation to the solution of sustainable development challenges. Being comprehensive, the SDGs present an opportunity for the development and implementation of business-led solutions and technologies to address a wide range of development challenges (GRI, UNGC, and WBSCD, 2015; Ohno et al., 2019) (Fig. 2). More specifically, a report produced by the Business and Sustainable Development Commission (BSDC, 2017) estimates that achievement of the SDGs by 2030 will create at least USD 12 trillion in market opportunities in four economic systems: food and agriculture; cities; energy and materials; and health and well-being. More than half of those SDG-related business opportunities will emerge in developing countries. The BSDC report also projects the creation of 380 million new jobs by 2030, 90% of them in developing countries (ibid.).

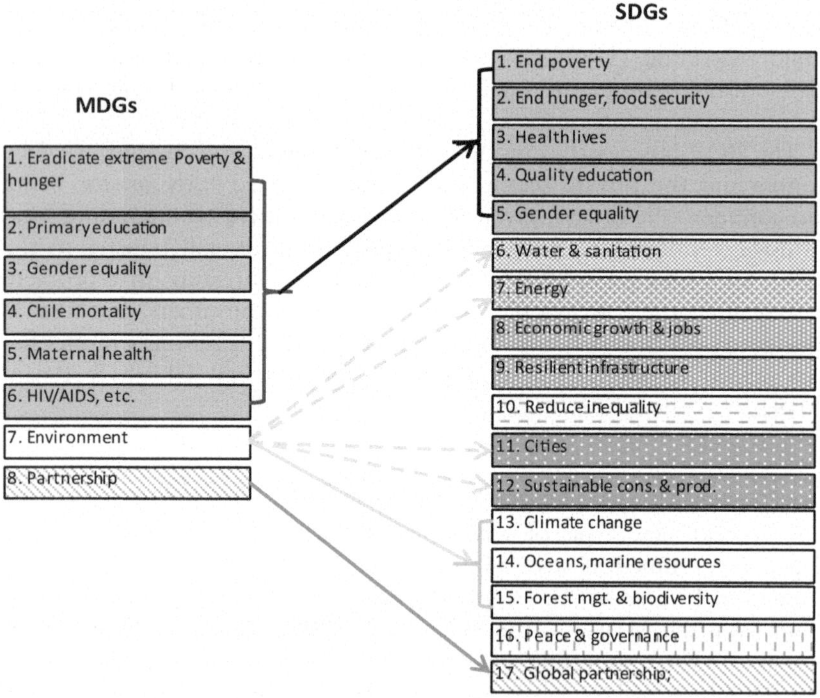

Fig. 2 Evolution from MDGs to SDGs (*Source* Ohno and Uesu [2020], based on information provided by JICA)

With the shift to the SDGs, the concept of BOP business has been broadened to address more comprehensive development challenges, and many donors have begun to actively engage in development partnerships for sustainable development. For example, since 2009, the German government's Federal Ministry for Economic Cooperation and Development (BMZ) has been supporting develoPPP.de,[3] a program supporting development partnership with the private sector, involving three agencies, including the German Agency for International Cooperation (GIZ). DeveloPPP.de gives financial and professional support (cost-sharing) to German and European companies and their subsidiaries in developing countries and emerging economies. The priority areas include food security; decent production and working conditions; employment; vocational training; environmental protection; conservation of natural resources; mitigation of the impact of climate change; and technology transfer.

USAID has had its own long-standing private sector partnership program, Global Development Alliance (GDA), which is not limited to the American private sector. Launched in 2001, GDA co-invests resources based on the principle of shared value. More recently, GDA established Development Innovation Ventures (DIV) for the testing of new ideas; for quick and inexpensive evidence gathering to determine what works and does not; and for provision of ongoing support for proven solutions.

Furthermore, such bilateral donor initiatives have gradually developed into multilateral platforms. A good example is Business Call to Action (BCtA), launched by the UN in 2008 with funding from bilateral donors including Australia's Department of Foreign Affairs and Trade (DFAT), the United Kingdom's former Department for International Development (DFID),[4] the Swedish International Development Cooperation (SIDA), and USAID. BCtA is directed by a secretariat operated by the United Nations Development Programme (UNDP). As of 2019, this multilateral platform is supported by over 200 companies from various sectors in more than 70 countries.

[3] The German government (BMZ) established "develoPPP.de program" in 2009 by restructuring the PPP Facility which was set up in 1999 as a separate fund for development partnerships with the private sector. Development partnerships aim to combine the innovative power of business with the resources, knowledge, and experience of development work.

[4] Under the Boris Johnson administration, DFID was merged with FCO (Foreign and Commonwealth Office) to form FCDO (Foreign, Commonwealth and Development Office) in September 2020.

2.3 *Corporate Social Responsibility in Value Chain Management*

The third new type of development partnership emphasizes responsible business. As MNCs increasingly invest in developing countries hindered by weak governance and low enforcement capacity, there is growing recognition among companies of the need to work in partnership with donor agencies and non-profit organizations (NPOs) to ensure social and environmental compliance in value chain management.

The recent expansion of GVCs provides opportunities for developing countries to participate in global markets and acquire knowledge and technology from foreign direct investment (FDI) companies. In the industrial sector, such participation occurs mostly at the entry level of the garment sector, which requires abundant low-wage labor with relatively low skills. In this regard, partnership with donor agencies can compensate to some extent for such deficiencies. In fact, when the grim conditions in garment industry sweatshops were made public, Western donors and CSOs moved to enhance development partnerships for the promotion of sustainable production and manufacturing. For instance, the International Labour Organization (ILO), together with the IFC, introduced the Decent Work Program in 2006 to improve labor standards and competitiveness in global supply chains, especially those related to labor-intensive industries in developing countries. The program assists companies to improve their practices through the application of core ILO labor standards and national labor laws, and the provision of technical assistance in areas such as worker-management cooperation, working conditions, and social dialogue. The program is supported by bilateral donors including Australia, GIZ, US, the Netherlands, the Danish International Development Agency (DANIDA), major apparel brands, CSOs, and the governments of developing countries.

The above examples demonstrate that development partnerships with the private sector have expanded significantly through bilateral and multilateral efforts and are now producing synergies in the form of multistakeholder global platforms. The scope of those partnerships has also broadened from the initial narrow focus on poverty reduction (typically, BOP business in the MDG era) to include economic upgrading and sustainability agenda, including environmental and social compliance (corresponding to the SDG era).

3 Evolution of Japanese Development Partnerships with the Private Sector

The external and domestic environment surrounding Japanese ODA and business has also changed significantly over the past decade. In response to accelerated outward FDI by Japanese firms, triggered by the 2008 financial crisis, the Japanese government began in 2011 to actively engage in the promotion of partnerships with the private sector as part of its ODA activities. Economic cooperation agencies such as the Ministry of Economy, Trade and Industry (METI), Japan External Trade Organization (JETRO), and the Association for Overseas Technical Cooperation and Sustainable Partnership (AOTS)[5] have long supported Japanese business, forging PPPs as the core of their operations. Today, even the Ministry of Foreign Affairs (MOFA) and JICA—which have previously demonstrated considerable reluctance to support individual firms and have rarely become involved in any business proposals presented to them—have started to use the ODA budget to support new private sector partnership activities, and to provide support for the overseas expansion of small- and medium-sized enterprises (SMEs). Another interesting development is that METI and related agencies, in an effort to broaden the interface between the private sector and development, have begun to encourage Japanese companies to promote the BOP business model.

Preoccupied with the hollowing out of domestic industry in Japan, the national government and local governments have until recently taken a cautious approach to supporting overseas expansion of SMEs. However, since 2010, there has been a new wave of internationalization, as growing numbers of SMEs have sought fresh opportunities abroad. Not only did those SMEs demonstrate a willingness to take on new risks; they also made those moves without the usual support of a parent company. In response, the government shifted its policy away from a guarded stance to active support of those pioneers (SME Agency, 2010).[6]

[5] AOTS supports industrial human resource development in developing countries, in partnership with the private sector and provides technical cooperation through training, expert dispatch, and other programs. The current AOTS, called HIDA (The Overseas Human Resources and Industry Development Association) during 2012–2017, was created in April 2012 through the merger of the Japan Overseas Development Corporation (JODC, charged with expert dispatch), the Association for Overseas Technical Scholarship (AOTS, charged with training), and other related structures within METI.

[6] A major turning point was the *White Paper on Small and Medium Sized Enterprise, 2010*, which discussed the need "to harness global economic development for Japan's

3.1 *From ODA to Development Cooperation*

In February 2015, the Japanese Cabinet adopted the Charter for Development Cooperation as a new set of guiding principles for navigating the age of the SDGs. Behind this latest revision is the realization that both Japan and the world are entering a new era.[7] The 2003 Charter was drawn up at a time when ODA flows from developed countries constituted the primary source of development finance. Today, private sector financial flows into developing countries far exceed ODA flows (MOFA, 2015). In addition, the impact of the Great East Japan Earthquake in March 2011 should not be underestimated. That disaster was so devastating and painful that the whole nation united to support recovery and reconstruction efforts in the Tohoku region. Alliances were forged among non-governmental organizations (NGOs), businesses, local communities, and the Self-Defense Forces, and innovative activities began to emerge, including social business and social investment to support people in the affected areas.

The key features of the 2015 Development Cooperation Charter are a strong emphasis on "quality growth" that is inclusive, sustainable, and resilient; and the promotion of private sector partnerships, including with Japanese business, to achieve that growth. The new title, Development Cooperation Charter, recognizes the changing landscape of international development, including the growing reach of the private sector and the increasing presence of emerging economies. The charter therefore speaks to not only developed countries, but also a range of other actors such as businesses, NGOs, local government, and emerging economies that are expected to use their particular strengths to cooperate in efforts toward the common goal of sustainable global development and development of individual countries. The new charter title also carries a sense of equal partnership based on cooperation for mutual benefit and two-way dialogue, and a move away from current conventional aid-based, one-sided donor-recipient relationships.

growth" and "to actively promote internationalization of SMEs in challenging conditions abroad" in order to further develop Japan's SMEs.

[7] Foreign Minister Kishida gave a speech at the Press Club on March 28, 2014, entitled "An Evolving ODA: For the World's Future and the Japan's Future." It is available at: http://www.mofa.go.jp/ic/ap_m/page3e_000169.html.

The motivation for JICA and MOFA to utilize ODA to progress private sector partnerships is linked to a number of different but related factors, particularly: (i) increased recognition of the importance of private capital for poverty reduction and sustainable growth in developing countries; (ii) the potential application of private sector technology, know-how and innovation, including that of Japanese companies, to the resolution of global challenges related to the environment, climate change, infectious diseases, food security, and energy supply; and (iii) the strong perception that such partnerships would be mutually beneficial for both developing countries and Japan. This line of thinking was already evident when the Democratic Party of Japan (DPJ) was in power.[8] However, when the Liberal Democratic Party (LDP) came to power at the end of 2012, this agenda was pushed further, as a high priority policy. Prime Minister Shinzo Abe unveiled Japan Revitalization Strategy: Japan is Back (Headquarters for Japan's Economic Revitalization, 2013), which designated global outreach as one of its three action plans. The global outreach action plan sets specific numerical targets for infrastructure exports and the overseas expansion of SMEs and leading medium-sized enterprises. It also envisages "the strategic use of economic cooperation."

3.2 Japan's New ODA Initiative for Business in Development

From around 2010, the Japanese government, together with JICA and JETRO, began to utilize ODA budget in the promotion of BOP business, and more recently broader sustainable business, to encourage the Japanese private sector (both large companies and SMEs) to engage in overseas business and utilize their technologies to solve development challenges. Two points are worthy of note. First, among these new initiatives, the BOP business scheme largely resembles those introduced by other donors from the early 2000s. Second, another scheme (launched in 2012) for supporting overseas expansion of Japanese SMEs is a unique initiative reflecting Japan-specific factors. Following the 2008 financial crisis, traditional Japanese-style keiretsu production relations crumbled,

[8] For example, the ODA Review compiled under the guidance of Minister of Foreign Affairs Katsuya Okada put forward the concept of "enlightened national interest"; the Government of Yoshihiko Noda announced the Rebirth of Japan: A Comprehensive Strategy in 2012. It is available at: https://www.cas.go.jp/jp/seisaku/npu/pdf/20120821/20120821_en.pdf.

and an increasing number of SMEs began to seek new opportunities overseas, making management decisions and taking risks to establish overseas production bases independently of large parent companies (Ohno, 2016)—which is why the initiative focuses on utilization of the knowledge and technologies of Japanese SMEs for the solution of development challenges.

Needless to say, economic cooperation agencies affiliated with METI, such as JETRO, AOTS, and the Organization for Small and Medium Enterprises and Regional Innovation (SMRJ), have long supported Japanese businesses across the board and are boosting measures to support the overseas expansion of Japanese SMEs by strengthening one-stop consultation services.[9] Those agencies also provide feasibility studies and hands-on support for export promotion and outward FDI. What is new is the engagement of MOFA and JICA in supporting the overseas expansion of Japanese SMEs. Table 1 provides examples of the newly introduced ODA modalities for support of partnerships with the Japanese private sector.

In September 2016, one year after the UN adoption of the SDGs, JICA formulated a position paper on the SDGs (JICA, 2016). Subsequently, in February 2017, JICA launched the SDGs Business Supporting Surveys (BSS)—a scheme for matching Japanese corporations and developing countries seeking knowledge and technology upgrading to meet development challenges beyond the scope of BOP business. The new matching scheme expands the scope of the existing BOP business scheme and highlights efforts to help establish partnerships for further industrial and economic development (Kobukura, 2018); since the launch of SDGs BSS in 2018, 53 proposals from the private sector have been adopted by JICA.

It should be noted that JICA's SDGs BSS integrates the two previous schemes, BOP business and support for overseas expansion of SMEs, into one framework (see Table 2 below). That encourages both corporations and SMEs to utilize their technologies, know-how and entrepreneurial

[9] In October 2010, METI established "Conference on Supporting SMEs in Overseas Business" to formulate strategy and strengthen the coordination functions to support SME overseas business expansion. At the implementation level, public agencies participating in this conference are expected to provide timely information to SMEs and make referrals to relevant agencies as necessary. Overseas, in selected countries, JETRO acts as a coordinating hub for the SME Overseas Expansion Platform, whose membership is drawn from the overseas offices of participating public agencies.

Table 1 ODA modalities to support partnerships with the private sector

Area of Support	*ODA Modality & Launch Date*	*Purpose*
Fact-finding surveys	Preparatory survey for infrastructure PPP projects [FY2009-, JICA) Preparatory survey for BOP business [FY2010-. JICA) – **Integrated into SDGs BSS (FY2018-)**	Support the design of infrastructure PPP projects that seek concessional loans or PSIF support Support the design of pro-poor, BOP business. **More recently, the scope has been broadened as SDGs business**
Pilot and dissemination	Pilot projects for disseminating private sector technologies for the socio-economic development of developing countries (FY2013- , JICA) – **Integrated into SDGs BSS (FY2018-)**	Support the introduction and dissemination of a company's products, technologies, and systems
Financial Cooperation	Private Sector Investment Finance (PSIF) (FY2012, after piloting from FY2010, JICA)	Provide financial support (loans and equity) to private sector projects that contribute to the socio-economic development of developing countries
Human resource development	Private-Sector Cooperation Volunteer System (FY2012-. JICA) Japan Centers (in six overseas locations, JICA)	Support global human resources development through the dispatch of private sector professionals as volunteers Train local business managers and workers focusing on Japanese business management
Targeted at SMEs Only		
Fact-finding surveys	Needs survey (FY2012-, MOFA: commissioned to JICA); Project formulation survey (FY2012-. MOFA: commissioned to JICA: FY2014-transferred to JICA)	Verify developmental needs for a company's technologies and products in developing countries, and support project formulation

(continued)

Table 1 (continued)

Area of Support	*ODA Modality & Launch Date*	*Purpose*
	Preliminary survey for preparing for SME outward FDI (FY2012-, JICA) – **Integrated into SDG BSS (FY2018-)**	Support information gathering and project formulation prior to outward FDI into developing countries
Pilot and dissemination	Feasibility study and pilot projects for disseminating SME's technologies in developing countries (FY2012-, JICA) – **Integrated into SDG BSS (FY2018-)**	Verify the suitability of SMEs' products and technologies in developing country and disseminating them
Grant aid	Non-project grant aid (FY2012-, MOFA)	Provide SMEs' products that contribute to the socio-economic development of developing countries, at their requests
Partnership with Local Governments		
Grass-roots technical cooperation	Grass-roots technical cooperation in partnership with local governments and for econonmic reactivation (FY2012 supplementary budget-, JICA)	Support the socio-economic development of developing countries, by mobilizing resources of local governments/communities (technologies, experience, human networks)

* The Japanese fiscal year runs from April 1 to March 31
Source Elaborated by the author, based on information provided by JICA and MOFA

Table 2 Outline of JICA-supported SDGs business supporting surveys

Program for Supporting SDGs Business and Japanese SMEs		
Small and Medium-sized Enterprise (SME) Partnership Promotion Survey	SDGs Business Model Formulation Survey with the Private Sector	SDGs Business Verification Survey with the Private Setor
Collection and analysis of basic data (several months to one year)	Examine feasibility of technologies, products, and knowhow to develop a business model proposal (several months to one year)	Verify technologies, products, and a business model to develop a business proposal (around 1–3 years)

Source JICA (2019)

spirit to address social and environmental challenges (in areas such as clean energy, smart cities and affordable housing), and to provide business solutions for the resolution of global issues and the meeting of various development challenges.

3.3 Science and Technology Research Partnership for Sustainable Development

In 2008, the Japanese government launched Science and Technology Research Partnership for Sustainable Development (SATREPS), an international joint research program with developing countries aimed at addressing global issues. SATREPS, the first initiative to integrate science and technology into Japanese development cooperation,[10] has three objectives: (i) to enhance international cooperation on science, technology and innovation (STI) between Japan and developing countries; (ii) to acquire new knowledge and technology that lead to the resolution of global issues and the advance of STI; and (iii) to boost self-reliant research and development capacity in developing countries through international joint research, the construction of sustainable research systems that can contribute to resolving issues, the coordination of networking among researchers, and the training of future human resources in developing countries and in Japan.

The target research fields fall into four categories: (i) environment and energy; (ii) bio-resources; (iii) natural disaster prevention and mitigation; and (iv) infectious disease control (see Table 3). Since 2008, a total of 157 projects have been initiated in 52 countries; as of June 25, 2020, there were 58 projects under implementation in 34 countries. By region, Asia has the highest number of projects (84 in 14 countries), followed by Africa (41 projects in 21 countries), Latin America and the Caribbean (23 projects in 9 countries), and other regions (9 projects in 8 countries).

[10] The SATREPS is jointly managed by the Japan Science and Technology Agency (JST) which provides competitive research funds for science and technology projects for academia, the Japan Agency for Medical Research and Development (AMED), which provides competitive research funds for medical research, and JICA, which assumes the implementation in the recipient countries. JST and AMED, supervised by the Ministry of Education, Culture, Sports, and Science and Technology (MEXT), are responsible for funding the research activities in Japan, while the research activities in developing countries are supported under the framework of ODA-funded technical cooperation.

Table 3 Research fields and areas of SATREPS

Research Fields	*Research Areas and Contributions to the SDGs*
Environment and Energy	Research contributing to the solution of global-scale environmental Issues *Contribution to SDGs—response to climate change, conservation of ecosystems and biodiversity, sustainable use of natural resources, and pollution prevention and control Research on realization of low carbon society and efficient use of energy * Contribution to SDGs - clean energy and climate action
Bio-resources	Research contributing to sustainable production and utilization of bio-resources *Contribution ot SDGs—food security, health promotion, nutrition Improvement, and sustainable agricutlure, forestry and fishery
Natural Disaster Prevention and Mitigation	Research on disaster prevention and mitigation towards social sustainability *Contribution to SDGs—analysis of disaster mechanisms, prior countermeasures, and disaster occurrence to post-disaster recovery and reconstruction processes
Infectious Disease Control	Research on measures to address infectious diseases control attuned to the needs of developing countries *Contribution to SDGs—response to global health challenges

Source Elaborated by the author, based on information provided by JST and AMED

The overarching goal of SATREPS is to realize "research outcomes of practical benefit to both local and global society" in two configurations: (i) the outcome is commercialized and prevails in the market; and (ii) the outcome is reflected in public services or has impact on public policy (MOFA, 2020). Several research fields in particular—sustainable use of natural resources and bio-resources—are expected to be compatible with commercialization (item (i) above), and Japanese firms have been involved in the SATREPS projects from an early stage of implementation (see Case 2 in Sect. 4).

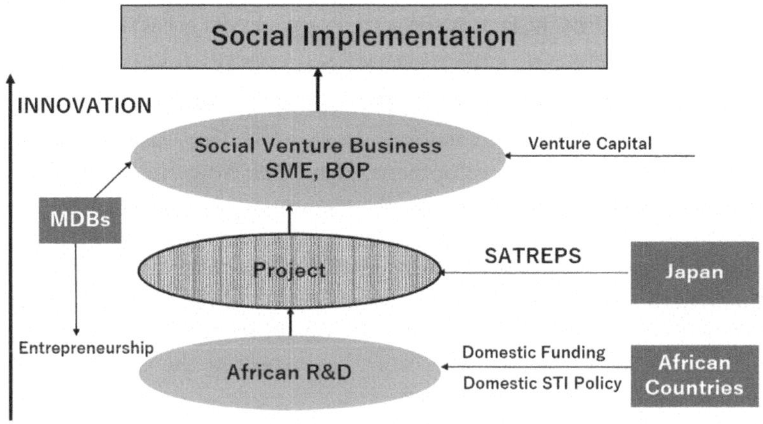

Fig. 3 Expected Role of SATREPS in Social Innovation in Africa. *Note* MDBs (Multilateral Development Banks), R&D (Research and Development) (*Source* Okaya [2012])

Later, in 2018, reflecting growing private sector interest in global issues, JST began assisting Japanese firms attracted to the use of the research outcomes to explore the development of SDGs business. Initially, three SATREPS projects were selected from among 120 programs, and by 2019, five projects had been identified, and calls had been made for private sector participation. The project themes include: (i) low-carbon economy in Thailand and Indonesia; (ii) environmental protection; (iii) bio-resources in Vietnam; (iv) water treatment in Nepal; and (v) earthquake risk mitigation in South Africa. However, third-party evaluation by MOFA (2020) revealed that the goal of commercialization of the research outcomes had been too ambitious to realize social impact, given the complex nature and the nascent stage of the joint research projects between Japan and developing countries.[11]

Figure 3 shows the SATREPS concept for advancement of social innovation and African development.

[11] We also benefitted from a comment from Prof. Jin SATO (a chief evaluator of the third-party evaluation team, also the author of chapter x in this book), via an online meeting on February 28, 2021.

4 Selected Cases of New Partnership Models of Development Cooperation

This section presents selected cases of the new partnership models of development cooperation explained above. The first case, an example of PPPs for sustainable development, describes how the addressing of environmental challenges using locally accumulated experience and technologies in combination with indigenously fostered social consciousness (Japanese *sanpo-yoshi*) has evolved into comprehensive environmental partnerships involving local governments, local SMEs, and the governments of Asian countries. The second case illustrates the way in which SATREPS scientific research outcomes have spurred the development of SDGs business in partner countries.

4.1 Case 1: Turning Local Environmental Challenges into PPPs for Sustainable Development

In the process of economic development, many countries confront environmental and social problems, and Japan was not an exception. Japan learned hard lessons from a number of negative experiences of local diseases caused by industrial pollution and waste disposal; and traffic congestion resulting from urbanization. Based on such experiences, some local governments—for example, in Kita-Kyusyu City and Yokohama City—are actively sharing their knowledge and expertise with emerging and developing countries in partnership with local companies and NPOs, to empower those countries to pursue economic development in a more sustainable manner. Moreover, it is worth mentioning that the indigenous Japanese notion of sustainable development (*sanpo-yoshi*, see Box 1) and CSR was widely shared among the Ohmi merchants active in the central Japan (Shiga Prefecture and surrounds) from the beginning of the eighteenth century onward.[12] In that context, the Lake Biwa Model can be regarded as a good example of comprehensive environmental partnerships among local government and local SMEs, for implementation in emerging and developing countries.

[12] With its geographical proximity to Kyoto and Osaka, the province of Ohmi (now the Shiga Prefecture) had been home to the itinerant merchants (Ohmi merchants). In addition, the abundant water supply from the Lake Biwa had much contributed to the economic development of central Japan (such as Kyoto and Osaka).

Box 1: *Sanpo-yoshi*—Traditional Japanese CSR

The term *sanpo-yoshi* means three-way satisfaction: good for the sellers (producers); good for the buyers (consumers); and good for the world. Suenaga (2019) observes that "the *sanpo-yoshi* emphasizes the importance of social consciousness and serving as a member of society" (p. 30), and that "the "good for the world" teaching of *sanpo-yoshi* demonstrates the premise that environmental preservation is absolutely essential, ... a hint to solve the dilemma faced by modern economics" (p. 31).

Those three principles became the code of the Ohmi itinerant merchants, several of whom later founded big commercial companies and MNCs such as Marubeni and Itochu. The *sanpo-yoshi* notion has been widely adopted by entities including local SMEs. For example, Hiyoshi Company, a local Shiga Prefecture SME specializing in water environment protection, has played a catalytic role in exporting the Lake Biwa Model to Asia. The company is also known for its adoption of the *sanpo-yoshi* principles and even extending it to *yonpo-yosh*i (four-way satisfaction), which also aims at goodness for the next generation.

Lake Biwa, which lies in Shiga Prefecture, has long been home to many manufacturers. The surrounding area is highly industrialized, with many factories operating between the Kansai and Tokai regions. Lake Biwa, the largest lake in Japan, is the water resource for a significant number of people and industries. Shiga Prefecture has taken measures to improve Biwa water quality since the 1970s, adopting advanced water treatment systems ahead of the rest of Japan, and promoting them throughout the prefecture. As a result, Shiga Prefecture boasts a very high penetration rate of systems for sewage treatment and advanced water treatment.

The Lake Biwa Model has been developed and consolidated over time, to the point where it is a unique public-private environmental platform for provision of a set of comprehensive actions for water conservation and economic development in the surrounding area. The model is built around local technologies and know-how accumulated among the residents and local communities, firms, research institutes, and the local government. Some companies in the area are collaborating with government agencies or other companies to develop new technologies and services to support further business expansion and development.

In March 2013, building on the Lake Biwa Model, with the aim of supporting the local private sector, Shiga Prefecture established the Shiga Water Environment Business Promotion Forum, a network aimed at developing new business projects through industry-government-academia collaboration (Shiga Prefecture, 2013). The forum provides the members (162 companies and organizations as of March 2018) with information about the latest trends in the water environment business; progressive approaches adopted by local companies; and various support measures. Shiga Prefecture operates the forum as a center for business matching to create new niche business projects and to find partners for joint development within and outside Japan.

The Lake Biwa Model is applicable not only to other lakes, but also to various water-related settings, such as seas, rivers, and industrial parks. Recently, it has been exported to China, Vietnam, and other Asian countries, with the assistance of JICA and other governmental funding schemes (explained in detail in Sect. 3). With a population of approximately 30,000, Cat Ba Island in the Ha Long Bay in Vietnam has developed flourishing fishing and tourism industries. Over the past ten years, however, the level of water pollutants increased due to the growing number of tourists and people living along the shore, causing deterioration of water quality, offensive odors, and a red tide outbreak. To improve the situation and promote "green growth," the local government applied the Lake Biwa Model, which led to enhanced environmental awareness among the local government, companies, and residents of Cat Ba Island, laying the foundation for environmental protection activities through industry academia-government collaboration.

4.2 Case 2: Linking SATREPS Outcomes to the Development of SDGs Business in Africa

As explained in the Sect. 4.1, recently, a new initiative has emerged to link the research outcomes of SATREPS projects to the development of SDGs business. Two cases in North Africa (Tunisia and Morocco) and one in Mozambique give insight into the way in which SATREPS projects have explored partnerships with the private sector from the early stages of project development. Both projects aim to make use of local plants cultivated in arid zones for industrialization.

The first case is the project entitled, "Valorization of Bio-resources based on Scientific Evidence in Semi- and Arid Land for Creation of New Industry" initiated in 2010 through a partnership between the University of Tsukuba in Japan and the Biotechnology Centre of Borj Cedria (CBBC) in Tunisia. This SATREPS project builds on the medicinal effects of local plants (e.g., polyphenol extracted from local olives for prevention of the onset of breast cancer; rosemary for anti-aging effect) that had been overlooked. The use of such local bio-resources, which are reflected in Islamic values and culture, is expected to offer huge potential for local value creation in the Maghreb and to open market opportunities in bordering regions including sub-Saharan Africa.

Moreover, the project has greatly benefited from: (i) solid development cooperation since the 1990s between Tunisia and Japan in the areas of higher education and STI (for instance, Japanese financial and technical cooperation to assist CBBC); and (ii) a long-term academic network among Japan and North African countries represented by the Alliance for Research on North Africa (ARENA), which has been operating at the University of Tsukuba since 2004 and expanded in 2018 to the Alliance for Research on the Mediterranean and North Africa. ARENA is interdisciplinary in nature, combining various research topics including bio-medical analysis; local socio-economic and cultural examination of factors related to the production area and producers; and marketing studies encompassing North Africa, Europe, and Japan. Joint research activities have been developed for some years in both Tunisia and Japan; these are critical for the consolidation and deepening of partnerships between public and private entities. Despite Tunisia's radical regime change in 2011, which led to political and economic instability in subsequent years, ARENA continued steadily organizing joint academic seminars, bringing together a wide range of researchers from Japan and North Africa. In addition, Japanese researchers have been proactive in attracting private sector interest so as to promote commercialization.

Building on the above-mentioned long-term multi-layered partnerships, the second phase of this SATREPS project (2016–2021) ran under the PPP framework linked to SDGs business and is expected to contribute to SDGs 3 and 15 (ARENA, n.d.). The project strengthened partnerships with the private sector and concluded joint research agreements with five companies in 2018. In 2019, Tsukuba University, CBBC, a Tunisian company (Herbes de Tunisie), a Japanese company (Nutrition Act), and JICA signed a Memorandum of Understanding. An effort was also made in Morocco to replicate the Tunisian experience; but Japanese researchers have faced a number of difficulties in concluding the agreement for PPP since its inception in 2016. Japanese and Moroccan universities and a local firm (Atlas Olive Oil Company) signed the first partnership agreement in October 2019. A Japanese SME and a local firm are expected to conclude (with the assistance of JICA's BOP and SDGs BSS) a partnership agreement to support the production and commercialization of argan oil, originated from local argan tree (ARENA, 2019).

The second case is related to the SATREPS project, "Sustainable Jatropha Biofuel Production in Mozambique Creating Environmentally Friendly Fuel from Land Unsuitable for Agriculture" (2010–2016). The project, aimed at the use of another local plant, jatropha, for industrialization, unfortunately missed its initial objective. But, it has actually expanded into social innovation, initiated by an environmental startup company. Initially, the production of jatropha—a biodiesel fuel plant cultivable on infertile land—was expected to offer an alternative approach to reducing CO2 emissions, and its development was widely encouraged in some developing countries, including Mozambique. In 2005, the government of Mozambique approved efforts toward the use of jatropha as an alternative fuel, and several studies were conducted to explore its development. Among Japanese public institutions, the cultivation of jatropha was initially tested in a commercialization study by the Global Environment Centre (GEC). Subsequently, SATREPS funded a group of researchers from the University of Tokyo, Kanazawa Institute of Technology, and Kurume University, in partnership with Eduardo Mondiane University in Mozambique under the above-mentioned project. That joint research is aimed at ensuring both economic viability and positive environmental impact, improving the lives of the local residents through the cultivation of jatropha and promotion of the use of byproduct, which will prevent further deforestation. Another component of the project is focused on the identification of technologies to ensure the quality and safety of the manufactured fuel, with a view to future industrialization in Mozambique and replication in other African countries.

During the implementation phase, however, it was necessary to revise the initial objectives of the project, particularly those related to the industrialization component, since global demand for biofuel had shrunk, prompting many larger commercial plants to leave Mozambique. In 2017, the government of Mozambique announced its decision to halt the development of jatropha as a potential source of biodiesel. An ex-post evaluation of SATREPS concluded that the research outcomes were "mixed" (JST, 2016), which left little room for local industrialization and few business opportunities for Mozambique and Japan. After the completion of SATREPS in 2016, the project was followed up in Inhambane Province under the title, "Small-scale Rural Electrification with

Jatropha Biofuel," under a grass-roots cooperation framework involving the Hokuriku branch of JICA in partnership with Kanazawa Institute of Technology (2017–2020).

Throughout that process, Nippon Biodiesel Fuel (NBF), a Japanese environmental startup founded in 2000, joined the above SATREPS project in 2011 and established a local company (Agro-Negócio para o Desenvolvimento de Moçambique, Lda: ADM) to explore the commercialization of jatropha in Capo Delgado in northern Mozambique. The New Energy and Industrial Technology Development Organization (NEDO) provided financial support for that commercialization from 2012 to 2013. According to NBF, about 7,000 farmers participated in the cultivation, with annual production of approximately 100 kL. After the completion of SATREPS, the startup continued its activities in the area and recognized that its business transactions had been constrained by lack of financial service. NBF then decided to expand the business and introduced an electronic voucher system in collaboration with Food and Agriculture Organization of the United Nations (FAO) in 2015, which made it possible to provide farmers with subsidies for the purchase of better agricultural inputs. The e-voucher is considered an ideal financial inclusion tool for familiarizing farmers with e-money systems and Information and Communication Technology (ICT) (GoJ, 2019). Now, several donors and private CSR have funded the implementation of a digital information platform (AgriNET) to give small-scale producers of jatropha access to market information and a variety of services (Iizuka, 2020).

This case has been hailed as a prominent success story of SDGs business among Japanese policymakers and practitioners. In 2020, JST encouraged joint research to promote ICT adoption and social innovation for Africa, as stipulated in a set of recommendations[13] by the Meeting of the Advisory Board for the Promotion of Science and Technology Diplomacy (JST, 2020) and promoted at the following the seventh Tokyo International Conference on African Development (TICAD 7) in 2019. However, at the time of writing, Capo Delgado has become a new epicenter of violent insurgency, and the residents are now fleeing the region en masse. Moreover, with the spread of COVID-19, little is known about the actual state of the digital information platform. Although the

[13] It stipulates the importance of 4 "Ss" (SDGs, STI, SATREPS, and Society 5.0) to promote the use of STI for social innovation and capacity building.

shift to SDGs business attracted considerable external funding and technical assistance, partner capacity building and STI promotion at the local level—core principles of SATREPS—seem to have fallen by the wayside in the process.

In sum, the Mozambique project, unlike the North African case, suffered from weak local leadership during SATREPS (JST, 2016) and has not yet been able to establish the necessary foundation for African research and development, specifically the conditions outlined in Fig. 2.

5 Implications of New Development Partnership Models and the Way Forward for Japan

In the concluding section, we consider implications of new development partnership models from two perspectives: (i) comparison with the traditional Japanese PPP model and new partnership requirements; and (ii) enhanced role of new partnerships in post-pandemic recovery. We then explore the way forward for Japan.

5.1 Comparison of the New Development Partnership and Traditional PPP Model

For over six decades, Japan has provided cooperation in East Asia as a combination of trade, investment, and ODA, successfully incorporating its partner countries into dynamic production networks that have contributed significantly to the economic progress of the region. This phenomenon is often referred to as the "East Asian miracle" (World Bank, 1993), and Japan's METI (2005) has dubbed that mix the Japan ODA Model. Japanese ODA has played a catalytic role in private sector development on a number of fronts: notably, it has upgraded productive infrastructure through concessional ODA loans; built institutions and systems to strengthen the functioning of the market economy; and developed human resources through technical cooperation. The resulting improved investment climate has attracted FDI from Japan and elsewhere, which has encouraged trade based on the import and export of raw materials and finished goods, in turn contributing to the development of partner countries (Watanabe & Miura, 2003).[14]

In this model, the main driver was the overseas expansion of large Japanese corporations and affiliated SMEs within a traditional Japanese production relationship. In contrast, the new wave of outward FDI regards developing countries and emerging economies not only as destinations for production bases, but also as important consumer markets. Moreover, the players are not only large corporations and their affiliates; a wide range of businesses are involved, including SMEs acting independently of big business. In this new era in which Asia has become both the center of global production and an important market, it is all the more important for companies investing abroad to build strong networks of local talent and organizations in order to succeed. Japan needs to move beyond traditional PPPs to form new style partnership with the private sector, incorporating both domestic and local stakeholders, as shown in Case 1 (the Lake Biwa Model). The need for that shift reflects not only the profit-making motive of Japanese companies, but also the need to adopt a developmental perspective.

To that end, the localization of business activities, such as research and development, planning, and distribution and sales, is indispensable. SATREPS is a new endeavor aimed at promoting joint research between developing countries and Japan, and at the practical application of research outcomes. At the same time, Case 2 (valorization of local plants in North Africa) shows that merely transplanting Japanese technologies may not be sufficient, and that local partners, such as local researchers, businesses, NGOs, and community organizations, play critical roles in such localization processes. The same holds true for SMEs expanding abroad. In general, small factories in Japan possess superior technology, but they face enormous structural constraints in terms of their ability to devise business strategy, draw up sales and marketing plans, and handle foreign languages. Thus, in cases where a Japanese SME plans to expand independently overseas, it needs to compensate for such resource constraints with local talent and organizational support.

[14] See also the *Interim Report of the Advisory Council on International Cooperation* (January 2008) chaired by Professor Toshio Watanabe. As a private advisory body to the Minister of Foreign Affairs, the Council was convened between March 2007 and February 2009 to discuss and make recommendations on basic policy on international cooperation from a diverse range of perspectives.

Compared to large corporations, Japanese SMEs need to build much deeper partnerships with local stakeholders in order to sustain their business activities (Ohno, 2015). Also, the framework for SATREPS shows the need for partnership with a wider range of actors in research and development—both domestically and externally—than the traditional ODA stakeholders.

In sum, to build the local networks described above, the new partnerships require much broader and deeper interaction with partner countries as well as stakeholders within Japan, far beyond the scope of the traditional PPP (based on a combination of investment, trade, and aid) that Japan has practiced until now.

5.2 Enhanced Role of New Partnership Models for Post-Pandemic Recovery

The COVID-19 pandemic has had broad effects on the entire world—not limited to the health sector, but impacting the whole economy and society. COVID-19 has slowed down the rate of global poverty reduction, and in fact reversed it for the first time in 20 years, posing substantial challenges regarding the achievement the SDGs by 2030 (World Bank, 2020). At the same time, there is an emerging global call for "building back better" in the post-pandemic recovery period by creating more sustainable, resilient, and inclusive societies.[15] Now the role of the private sector is even more important, and new partnership models have even greater potential to contribute.

In particular, COVID-19 has accelerated the ongoing digital transformation. There is rapid development of a wide variety of new technologies (often referred to as "Corona-Tech") for the solution—especially by startups—of enormous problems caused by the pandemic, and the addressing of already existing development challenges. It is important to support SMEs and startup firms that possess advanced technologies and innovative ideas, to empower them to make useful contributions to the better recovery and the achievement of the SDGs. At the same time, as shown

[15] The UN Secretary-General António Guterres's statement in his International Mother Earth Day message (April 22, 2020) that "We need to turn the recovery into a real opportunity to do things right for the future." https://www.un.org/en/un-coronavirus-communications-team/un-urges-countries-%E2%80%98build-back-better%E2%80%99.

by the two SATREPS projects in Case 2, patience and long-term perspectives are needed to nurture development partnerships with the private sector, and ongoing effort must be made to identify appropriate support measures and instruments for SDGs business and improve them.

References

ARENA. (n.d.). *About the arena* (in Japanese). https://arenatsukuba.wordpress.com/center/purpose/. Accessed 15 March 2021.

Alliance for Research on the Mediterranean and North Africa, University of Tsukuba (ARENA). (2019). *News: First partnership agreement in Morocco* (in Japanese). https://arenatsukuba.wordpress.com/2019/11/14/. Accessed 15 December 2020.

Barrientos, S., Gereffi, G., & Rossi, A. (2011). Economic and social upgrading in global production networks: A new paradigm for a changing world. *International Labour Review, 150*(3–4), 319–340. https://doi.org/10.1111/j.1564-913X.2011.00119.x. Accessed 15 December 2020.

Business and Sustainable Development Commission (BSDC). (2017). *Better business, better world*. https://www.unglobalcompact.org/library/5051. Accessed 14 May 2021.

Ghosh, S., & Rajan, J. (2019). The business case for SDGs: An analysis of inclusive business models in emerging economies. *International Journal of Sustainable Development and World Ecology, 26*(4), 344–353. https://www.tandfonline.com/doi/full/10.1080/13504509.2019.1591539. Accessed 15 December 2020.

Global Reporting Initiative (GRI), United Nations Global Compact (UNGC), World Business Council for Sustainable Development (WBSCD). (2015). *SDG compass: The guide for business action on the SDGs*.

Government of Japan (GoJ). (2019, July 10). *Transforming Africa through information and communication technologies*. https://www.japan.go.jp/sustainable_future/japaninafrica/articles/en_09.html. Accessed 20 February 2021.

Groɜe-Puppendahl, S., Byiers, B., & Bilal, S. (2016). *Beyond aid in private sector engagement: A mapping of the opportunities and challenges of development and commercially-oriented public support to private sector engagement* (ECDPM Discussion Paper No. 187). European Centre for Development Policy Management (ECDPM). https://ecdpm.org/publications/beyond-aid-private-sector-engagement. Accessed 15 December 2020.

Hammond, A. L., Kramer, W. J., Katz, R. S., & Tran, J. T. (2007). *The next 4 billion: Market size and business strategy at the base of the pyramid*. World Bank Group. http://documents.worldbank.org/curated/en/779321468175731439/The-next-4-billion-market-size-and-business-strategy-at-the-base-of-the-pyramid. Accessed 15 December 2020.

Headquarters for Japan's Economic Revitalization. (2013). *Japan revitalization strategy: Japan is back.* Cabinet decision on 14 June 2013. http://www.kantei.go.jp/jp/singi/keizaisaisei/pdf/en_saikou_jpn_hon.pdf. Accessed 15 December 2020.

Iizuka, M. (2020, June). *Building partnership and networks for acceleration and expansion of the pilot programme: Role of think tanks/academia, foundations and private sector—Some snapshots from private sector*. Presented at the Workshop on Developing STI Partnerships for the Sustainable Development. http://ris.org.in/sti/files/pdf/ppt/day%202/200629_DAY2_SessionIV_IIZUKA.pdf. Accessed 15 March 2021.

Japan International Cooperation Agency (JICA) (2016, September 8). *Toward achieving sustainable development goals (SDGs)*. JICA's Position Paper on the SDGs. 2016. https://www.jica.go.jp/aboutoda/sdgs/ku57pq00002e2b2a-att/JICA_torikumi_e.pdf. Accessed 15 December 2020.

JICA. (2019). *SDGs Business supporting surveys*. Private Sector Partnership and Finance Department, JICA. https://www.jica.go.jp/priv_partner/case/ku57pq00002avzny-att/ind_summary_en_201907.pdf. Accessed 14 May 2021.

JST. (n.d.). *SATREPS, Research fields and areas*. https://www.jst.go.jp/global/english/area_of_research.html. Accessed 15 December 2020.

Japan Science and Technology Agency (JST). (2016). *SATREPS kenkyu kadaibetsu shuryoji hyouka houkokusyo* [Ex-Post Evaluation of SATREPS in Mozambique]. https://www.jst.go.jp/global/hyouka/pdf/h2201_mozambique_terminal-evaluation-report.pdf. Accessed 3 March 2021.

JST. (2020). *SATREPS kokusai gijyutsu kyouryoku proguramu koubo youryo* [Application for SATREPS International Collaborative Research Program]. https://www.jst.go.jp/global/pdf/koubo2021.pdf. Accessed 8 April 2021.

Kobukura, K. (2018). *Strengthening partnerships and collaboration with all stakeholders is key to achieve all SDGs*. Agenda 2030 and public private partnership (Discussion Paper). Friends of Europe. https://www.friendsofeurope.org/insights/agenda-2030-public-private-partnership/. Accessed 15 December 2020.

Ministry of Economy, Trade and Industry (METI). (2005). *Waga kuni keizai kyoryoku no seiko keiken o fumaeta JAPAN ODA moderu no suishin* [The Promotion of the Japan: ODA model based on the success of Japan's Economic Cooperation]. Interim Report. Economic Cooperation Group, Subcommittee on Economic Cooperation, Industrial Structure Council. Tokyo.

Ministry of Foreign Affairs (MOFA). (2015, February 10). *Cabinet decision on the development cooperation charter (provisional translation)*. http://www.mofa.go.jp/files/000067701.pdf. Accessed 15 December 2020.

MOFA. (2020). *Evaluation of the SATREPS program: Science and technology research partnership for sustainable development* (Third Party Evaluation

Report 2019). Ministry of Foreign Affairs of Japan. https://www.mofa.go.jp/policy/oda/evaluation/FY2019/pdfs/satreps.pdf. Accessed 1 March 2021.

Ohno, I. (Ed.). (2015). *Machi koba kara ajia no gurobaru kigyo e: chūshokigyo no kaigai shinshutsu senryaku to shiensaku* [From a small factory in Japan to a global firm in Asia: SMEs' overseas expansion strategy and policy support]. Chuokeizai-sha.

Ohno, I. (2016). Network-based development cooperation as a way forward for Japan. In A. Asplund & M. Söderberg (Eds.), *Japanese development cooperation: The making of an aid architecture pivoting to Asia* (pp. 159–176). Routledge.

Ohno, I., Konya, K., Shiga, H., Murillo, F., & Charvet, E. (2019). *Scaling up business impact on the SDGs*. Policy brief for T20 Japan 2019. Task Force on 2030 Agenda for Sustainable Development. JICA-Research Institute.

Ohno, I., & Uesu, S. (2020). New partnership between business and development: A case study of Ethiopia's industrial development. In S. T. Otsubo & C. S. Otchia (Eds.), *Designing integrated industrial policies volume II: For inclusive development in Africa and Asia* (pp. 203–236). Routledge.

Okaya, S. (2012, April 1–3). *Tackling global challenges: Science and Technology Research Partnership for Sustainable Development (SATREPS)*. Presented at the first Africa Forum on Science, Technology and Innovation (STI): For Youth Employment, Human Capital Development and Inclusive Growth held in Nairobi, Kenya. https://www.slideserve.com/kipling/shig-okaya-japan-science-and-technology-agency-jst-1-april-2012-nairobi. Accessed 27 February 2021.

Porter, M. E., & Kramer, M. R. (2011, January-February). The big idea: creating shared value. *Harvard Business Review* (pp. 2–17).

Prahalad, C. K. (2004). *The fortune at the bottom of the pyramid: Eradicating poverty through profits*. Wharton School Publishing.

Prahalad, C. K., & Hart, S. L. (1999, August). *Strategies for the bottom of the pyramid: creating sustainable development* (Unpublished draft paper). http://pdf.wri.org/2001summit_hartarticle.pdf. Accessed 15 December 2020.

Shiga Prefecture. (2013). *About the Shiga water environment business promotion forum*. https://www.pref.shiga.lg.jp/mizukankyobusiness/english/index.html. Accessed 15 December 2020.

Shimomura, Y., & Wang, P. (2013). The evolution of aid, investment, trade synthesis. In J. Sato & Y. Shimomura (Eds.), *The rise of Asian donors: Japan's impact on the evolution of emerging donors* (pp. 114–132). Routledge.

Small and Medium Enterprise Agency (SME Agency). (2010). *Chushokigyo hakusho 2010 nenban: pinchi o norikoete* [2010 White paper on small and medium enterprises in Japan: Pulling through the crisis]. METI.

Suenaga, K. (2019). *The story of Japan's Ohmi merchants: The precept of Sanpo-yoshi*. Japan Publishing Industry Foundation for Culture (JPIC).

Watanabe, T., & Miura, Y. (2003). *ODA (seifu kaihatsu enjo): nihon ni nani ga dekiruka* [ODA (Official Development Assistance): What Japan can do]. Chuokoron-shinsya.
World Bank. (1993). *The East Asian miracle: Economic growth and public policy* (World Bank Policy Research Report). Oxford University Press.
World Bank Group. (2020). *Poverty and shared prosperity 2020: Reversals of fortune*. The World Bank.

CHAPTER 8

Science, Technology, and Innovation in Sustainable Development Cooperation: Theories and Practices in South Korea

Kyung Ryul Park

1 Introduction

This chapter discusses South Korea's development cooperation policy with a particular focus on science, technology, and innovation (STI) and examines Korea's unique opportunities and challenges. STI have been recognized as key areas in achieving development. There have been a significant number of scholarly works examining how STI promote development, drawing upon various fields including economics, development studies, science, and technology policy. Earlier literature has discussed the effects of STI on economic growth, suggesting that STI developments increase productivity and efficiency as well as hasten societal progress.

K. R. Park (✉)
Korea Advanced Institute of Science and Technology (KAIST), Daejeon, South Korea
e-mail: park.kr@kaist.ac.kr

H.-j. Kwon et al. (eds.), *International Development Cooperation of Japan and South Korea*,
https://doi.org/10.1007/978-981-16-4601-0_8

Building on these, more recently the focus has been on how STI in sustainable development not only require appropriate research and development (R&D) strategies, infrastructure, and capabilities to enhance innovation, but also the capacity of individuals, universities, and private sector stakeholders. In parallel to this development, in the international development arena, governments and international development agencies have developed agenda settings and policies for STI-driven development initiatives. Following the recent establishment of the United Nations Sustainable Development Goals (SDGs), the Technology Facilitation Mechanism (TFM) and STI roadmaps and action plans have been widely implemented.

STI policies have played a significant role in the socio-economic transformation of a developing country. This is particularly true in Korea. One of Korea's outward-looking strategies in the previous developmental stage was to facilitate foreign capital investment and technology assistance while constantly cultivating the domestic capacity for economic development (Nam, 1997). During the industrialization period of Korea from the 1960s to 1980s, key science and technology policies in this catch-up phase include promoting technology transfer from advanced countries and international development agencies; importing knowledge to build domestic capacity; and supporting conglomerates to strengthen industrial capacity (Nam, 1997). Furthermore, in the post-industrial era, the emphasis of national STI policy has been shifting from solely economic growth toward achieving inclusiveness and sustainable prosperity. STI policy began to play an active role in tackling societal problems and promoting health care and environmental sustainability with the consideration of improving the quality of citizens' life.

STI became increasingly important in Korea's official development assistance (ODA), since it joined the Organisation for Economic Co-operation and Development (OECD) Development Assistance Committee (DAC). The government of Korea enhanced domestic technological capacity by promoting efficient technology policies and engaging the private sector as a major force for improving national capability. However, the role of STI in development cooperation is neither theoretically well understood academically nor strategically implemented in Korea's ODA policy process. At the policy and institutional level, STI has not been fully articulated in Korea's ODA, while building tensions between STI-focused ministries and ODA-related ministries. The design and implementation of an STI-focused ODA require engaging multiple stakeholders with an appropriate governance structure, reflecting domain knowledge and specific development sectors in line with national priorities. In enabling STI initiatives through policy change and institutional reforms, stakeholders should consider trade-offs to be made as well as synergies to be realized. Another important challenge is the lack of a common definition of science and technology ODA due to its cross-cutting nature, while this cross-cutting nature of STI makes it an important factor in reaching nearly all the ODA as well as the SDGs. Developing a statistical measurement of STI-related development cooperation remains a major challenge to be solved for better development effectiveness.

Building on the earlier theoretical and empirical discussions of international development cooperation, this chapter focuses on the role of STI in development cooperation and investigates policy and practices of South Korea. First, this chapter reviews the theoretical underpinnings of the relationship between STI and development. Second, building on a brief history of Korea's STI-related ODA, this chapter investigates how South Korea utilizes science and technology in development cooperation and identifies opportunities and existing institutional and policy challenges. Third, it identifies emerging issues and discusses recent policy change by examining the implementation of the Korean-SDGs and effects of the COVID-19 pandemic.

2 Theorizing STI and Development

2.1 *Defining STI in Development*

STI policy as an integral component of national development strategies is widely recognized as a key enabler for economic growth and sustainable development (Miedzinski et al., 2019; IATT, 2020). The notion of STI has evolved over the last 50 years as one of enabling components in economic growth. Growing recognition of the importance of STI has resulted in diverse policy initiatives in the field of international development including those in the OECD and advanced countries (Soete, 2019). However, drawing a boundary of STI from a development perspective is not easy. One of the most commonly used conceptual terms in the discussion of STI and development is technological innovation. Major leading economic growth paradigms including the neo-classical Solow model, the Schumpeterian, and the endogenous growth theory have attempted to investigate the effects of technological progress and productivity on development (Furman et al., 2002; Rostow, 1960; Solow, 1956). Notably, Romer's product-variety model suggests technological innovation results in productivity growth while creating diversities of products (Romer, 1990). He argued that technological innovation is not an exogenous byproduct of scientific research, rather, sheds light on the importance of government R&D policy and intellectual property rights in endogenous growth. Schot and Steinmueller (2018) suggest that STI can be understood in terms of three framings as they evolved over the past decades. The first framing was identified as beginning with the institutionalization and legitimization of government support for STI in the era of 'modernization.' It focuses on innovation for growth and is mainly directed toward mass production and consumption. The second framing emerged in the 1980s with globalization and focused on competitiveness constructed by the national innovation systems for knowledge creation. The third and emerging framing is linked to contemporary social and environmental challenges including the latest global agenda—the SDGs—and calls for transformative change in a country. This framing has become clearer in recent years. Developing countries may be able to build on the experience from other countries to catch up and generate accelerated development by leapfrogging (Juma et al., 2005). Although some developing countries have managed to significantly improve their STI capacity, many continue to struggle to meet basic needs such as adequate health, education, electricity, and information and communication technologies (ICTs) infrastructure. Not only is there a remaining technology gap between countries, but there are also significant differences within demographics. Among the diverse forms of STI, ICTs have been increasingly important and considered one of the most powerful tools that can contribute to not only progressive but also disruptive transformation in development (Avgerou, 2003; Heeks, 2008). Notably, scholarly works in the field of ICT for development (ICT4D) have explored incremental and disruptive mechanism of ICT-enabled transformation in developing countries (Avgerou & Walsham, 2000; Heeks, 2008; Sahay, 2017). ICTs have created new avenues for making development projects more transparent, cost-effective, and engaging to development partners and citizens in developing countries (Wittemyer et al., 2014). However, implementing new technologies do not simply result in economic and social development for all and can often amplify the existing inequality in the local context (Avgerou, 2010).

2.2 Technical Assistance and Technology Transfer

Technical assistance was the earliest form of science and technology related ODA to developing countries by donor countries and international organizations. The European Recovery Program (ERP) after the Second World War, commonly called the Marshall Plan, was a financial aid program sponsored by the United States that also provided technical assistance. Although it is estimated that only less than 1% of all ERP aid was spent on technical assistance, its effect was significant in promoting industrial and agricultural productivity in the recipient countries (Tarnoff, 2018). The foci of scholarly works also include not only the industrial and economic impact of technical assistance (Gamser, 1988; Godfrey et al., 2002) but also its effect on socio-political capacity and democracy (Gibson et al., 2015). Tracing the evolution of technical assistance, Wilson (2007) presents a conceptual link between technical assistance, knowledge management, and national innovation systems and calls for more cooperative learning in STI-related development initiatives.

Another topic that has attracted attention from practitioners in development policy and academics is technology transfer (Correa, 1994). It is recognized as the main determinant in the economic growth of developing countries (Glass & Saggi, 1998; Reddy & Zhao, 1990). Compared to technical assistance, technology transfer refers to a wide range of international cooperation between governments and also firms through a variety of modalities including ODA, foreign direct investment (FDI), firm acquisitions, licensing agreements, and joint ventures. Scholarly works have been engaged in diverse development sectors such as in the field of climate change (Forsyth, 2007; Karakosta et al., 2010), clean development mechanism (CDM) (Murphy et al., 2015), renewable energy (Wilkins, 2002), and more recently, ICTs (Avgerou, 2003; Baark & Heeks, 1999). In the process of technology transfer at the national as well as international level, socio-political factors and complex dynamics of local institutionalization are also discussed in scholarly debates (Guston, 1999; Oda, 1991). Analyzing the technology transfer patterns from the United States and Japan to South Korea during the industrialization period, Hahm et al. (1994) also investigate its dynamics and suggest a shift of focus on the host developing country and the contextualization of the domestic process of technology implementation. In addition to this perspective on technology transfer from the recipient side, Kapur (2001) provides a unique contribution and opens up a debate to theoretically discuss the complex relationship between technology transfer, 'brain drain,' and migration based on the human development approach.

2.3 *Increasing Role of Innovation in Development*

Innovation is commonly defined as 'a new or improved product or process (or a combination thereof) that differs significantly from the unit's previous products or processes and that has been made available to potential users (product) or brought into use by the unit (process)' by the Oslo Manual (OECD and Eurostat, 2018). This definition comprises more than just the research and development phase, as scholars and practitioners aimed to build capacity or change the work approach can also be taken into account. The concept of national innovation systems (NIS) has been widely adopted by both academics and practitioners to account for complex dynamics between diverse stakeholders from a more holistic socio-technical perspective (Nelson, 1993; Sharif, 2006). Criticizing the lack of attention to technological innovation in neoclassical growth models, Nelson (1993) notably suggests that the science and technology driven national innovation system is crucial and needs to be understood as a part of a large institutional system composed of different sectors including government, academia, and industry. The concept of NIS suggests a policy framework that less developed countries can apply to their development strategies, which would need diverse institutional interactions in local innovation systems.

Development agencies and international organizations have also recognized the key role of innovation system for development and their responsibility to increase efforts to support R&D (OECD, 1996; World Bank, 2011). To measure innovation, worldwide patent data has mainly been used as a means in much academic research. The World Intellectual Property Organization (WIPO) leads the development of an effective international intellectual property (IP) system. The Addis Ababa Action Agenda (AAAA) endorsed by the UN member states in 2015 also stresses the importance of such efforts to stimulate a culture of incentivized innovation. Measuring the expenditure of innovation remains a challenge, as firms do not specify this in their financial accounts, and countries do not effectively compile such statistics (Furman et al., 2002). Besides, it can be difficult to differentiate between development activities that support innovation and activities that are innovative in the way they support development. Although DAC's systems were not set up to measure innovation expenditure, there has been progress made in the recognition of national innovation systems and R&D as potential new approaches to measuring innovation.

In 2015, countries spent on average 1.7% of their GDP on R&D (Ericsson & Mealy, 2019). However, there are large differences between individual states. While higher-income countries already spent an average of 2.4% on R&D and continue to increase investments, lower-income countries remain static with only 0.4% of their GDP being used for R&D (Ericsson & Mealy, 2019). With no STI strategies in place, these countries miss out on the high potential for economic growth. With economic growth, R&D tends to become increasingly privatized, whereas in low-income countries funding comes almost entirely from the government or international sources. Although even in most high-income countries the government remains the main funder for R&D, Korea and Japan received approximately 75 and 78%, respectively, of their R&D funding from businesses in 2016, making them worldwide frontrunners in the privatization of R&D. It is interesting to note that the remaining portion of the funding in Korea is predominantly sourced by the government, whereas in Japan higher education has a significant share (roughly 5%) as well. Investments in R&D for STI have been considered an established means toward robust economies and sustainable development. However, critics argue that higher R&D spending does not necessarily mean greater innovation. More investments in developing countries are required to further develop infrastructure and technological capacity as well as to finally leap-frog development. Notably, expanding access to ICTs and more importantly contextual considerations for local innovation are expected to increase connectivity and knowledge transfer between development partners.

2.4 *Institutionalizing STI in Sustainable Development Goals*

The SDGs are designed to tackle key barriers to sustainable development by ending poverty, protecting the planet, and ensuring prosperity for all. In comparison with the Millennium Development Goals (MDGs) targeting 2015, which paved the fight against poverty in developing countries at the beginning of the century, the SDGs have a broader scope (Fukuda-parr & Muchhala, 2019). The SDGs framework is made up of 17 goals, 169 targets, and 304 indicators to monitor the progress of the goals and provide accountability for the implementation of the SDGs. The means of implementation (MoI) of the UN 2030 Agenda specify finance, technology, capacity building, trade, policy and institutional coherence, multi-stakeholder partnerships and data-driven monitoring and accountability as key means of implementation to achieve the SDGs and highlight STI as the central tool for SDG implementation (Walsh et al., 2020). Compared to the previous agenda in the field of international development including the MDGs and the series of Aid Effectiveness agendas, the adoption of the SDGs recognized the importance of the cross-cutting qualities of STI in global sustainable development.

On the way to the establishment of the SDGs as a global norm, in 2014, the United Nations Interagency Working Group on a Technology Facilitation Mechanism (TFM) identified opportunities to collectively achieve more significant impacts by mapping all relevant STI initiatives, stimulating cooperation, and sharing information (United Nations, 2018). The TFM was set up to facilitate multi-stakeholder cooperation and collaboration toward access to STI through the sharing of information, experiences, best practices, and policy advice among member states to achieve the SDGs. The TFM also calls for the promotion of development, transfer, and dissemination of STI to developing countries, as well as capacity development in STI to help them achieve the SDGs (Walsh et al., 2020). New mechanisms to support countries' STI capacities were put in motion through the Third International Conference on Financing for Development (F4D) in Addis Ababa in July 2015 and later in the United Nations Sustainable Development Summit in September in New York. Notably, in the Addis Ababa Action Agenda (AAAA) endorsed by the UN member states, the TFM was established to support the SDGs, which were later officially adopted in the United Nation General Assembly. It was institutionally formalized and outlined in paragraph 123 and 124 of the Addis Ababa Action Agenda and paragraph 70 of the Post-2015 Development Agenda Outcome Document.

> We decide to establish a Technology Facilitation Mechanism. The mechanism will be launched at the United Nations summit for the adoption of the post-2015 development agenda in order to support the Sustainable Development Goals.
>
> - We decide that the Technology Facilitation Mechanism will be based on a multistakeholder collaboration between member states, civil society, the private sector, the scientific community, United Nations entities and other stakeholders and will be composed of a

United Nations Inter-Agency Task Team on Science, Technology and Innovation for the Sustainable Development Goals, a collaborative multi-stakeholder forum on science, technology and innovation for the sustainable development goals and an online platform.

(*Source*: Addis Ababa Action Agenda, Paragraph 123, pp. 55, 2015)

While defining the 2030 agenda at the UN Conference on Sustainable Development in 2015, member states placed more emphasis on STI and identified options for TFM that would help national governments to reach 'the future we want.' SDG 9 declares that technological progress is the foundation of efforts to achieve sustainable industrial and environmental development: "[...] Without technology and innovation, industrialization will not happen, and without industrialization, development will not happen." SDG 17 stresses the need for action in partnerships between governments, the private sector, and civil society—the stakeholders of a national system of innovation. SDG 17 explains that urgent action is needed to mobilize, private resources toward long-term investments addressing issues such as sustainable energy, infrastructure, and transport, as well as ICTs. Policymakers will need to set a clear direction and review policy frameworks, regulations, and incentive structures.

3 Korea's Development Cooperation with STI

3.1 Emergence of STI in Korea's ODA Policy

Korea joined the OECD Development Assistance Committee (DAC) in 2010. Korea's accession to the DAC was significant in several ways. First, the government played a leading role with the objective of heightening Korea's national status as an emerging donor country. Second, in addition to the endorsement of the Paris Declaration on Aid Effectiveness in 2005, the membership in DAC was meaningful for the Korean government to comply with various norms of international development cooperation such as the commitment to enhance development effectiveness and increase ODA and untied aid. Finally, building on the confidence of being transformed from one of the poorest countries to a DAC member, the Korean government has tried to establish a Korean ODA model to share development experiences and knowledge with developing countries

(Korea Institute for International Economic Policy, 2015; Lee, 2011; Yim, 2015).

Recognizing the importance of science and technology as well as the roles of government research institutes in the stage of Korea's economic development, STI has gradually attracted attention as one of the competitive strengths in Korea's ODA. On the ODA policy side, in 2011, the then Ministry of Education, Science, and Technology announced the 'national strategy of science, technology and education ODA for enhancing developing countries' capacity and supporting sustainable development.' The suggested main focuses of the government were (1) increasing ODA and strategic support, (2) enhancing developing countries' capacities through a Korean ODA model, (3) improving ODA to increase development efficiency, and (4) developing a foundation for STI-based ODA for the first time (Ministry of Education, Science, and Technology, 2011). Apart from these efforts that have been driven by STI-focused ministries including the Ministry of Science and ICT, the overall ODA strategies are discussed under the leadership of the Ministry of Foreign Affairs and the Prime Ministerial Committee of International Development Cooperation.

3.2 *Prioritization of ICTs*

Since Korea joined the OECD DAC, Korea has been considered one of the top providers of ICT ODA including e-government program, ICT education, and infrastructure. As ICT also has a cross-cutting nature, it may not be easy to disentangle technological components from the ICT ODA program. However, as ICT has been one of the top priorities in Korea's ODA program, it is meaningful to discuss its trends and policy change in the past years by investigating the size of ICT ODA compared to the total ODA based on the Korea's Annual ODA Plan which is determined by the International Development Cooperation Committee. KISDI (2018) and Yoo & Yoo (2019) provide the latest overview of ICT ODA and identify the access-related 'pure' ICT ODA in Korea. With the sole purpose of enhancing ICT infrastructure, pure ICT ODA accounted for 242.9 billion Korean Won (KRW) (2.3% of the total ODA) during the four years from 2015 to 2018 covered 10.5 trillion KRW (KISDI, 2018). However, ICT ODA can be defined more broadly to include all the ODA that incorporates ICT components, since it can contribute to various sectors in different ways due to its cross-cutting characteristic.

While there were fluctuations in the size of pure ICT ODA over the four years, the study shows that ICT ODA in general continued to grow from 4.8% in 2016, to 8.7% in 2017, and 8.8% in 2018. This shows that the trend in ICT ODA is focused on the convergence of ICT with various fields other than the pure ICT ODA as shown in Table 1.

The total sum of four years shows that ICT and public administration accounted for the largest share in ICT ODA (254.2 billion KRW, 32.1% of the total ICT ODA), followed by pure ICT ODA (242.9 billion KRW, 30.7%), ICT and economic–industrial infrastructure (130.6 billion KRW, 16.5%), ICT and education (78.3 billion KRW, 9.9%), ICT and environment, energy (40 billion KRW, 5.1%), and ICT and agriculture, forestry, and fisheries (1.7%) (KISDI, 2018). This shows the dominance of e-government related projects in ICT ODA in Korea. Notably, there were growing trends in ICT and education as well as ICT and economic and industrial infrastructure—only accounted for 3.8% of the total ICT ODA in 2015, but rapidly grew to account for 29.2% in 2018, the largest share in 2018. On the other hand, there is a significant decrease in pure ICT ODA, which accounted for 48.2% of the total ICT ODA in 2016 but dropped to 23.3% in 2017 (KISDI, 2018). This implies that while infrastructure and ICT and public administration (e-government) projects were at the center of ICT ODA in the past, ICT ODA is now expanding its portfolio to other various fields as well.

Table 2 shows the ICT ODA provided to different regions. Asia accounted for the largest portion with 37.6% (297.8 billion KRW) of the total ICT ODA, followed by Africa with 30% (213.5 billion KRW) and the Middle East/CIS with 17.2% (136.4 billion KRW) during the four years. However, while Asia accounted for 53% of the total ICT ODA in 2015, it was reduced to 40% in 2016 and 31.1% in 2017. In contrast, ICT ODA in the Middle East/CIS has steadily increased from 6.1% in 2015 to 10.1% in 2016, 18.7% in 2017, and 27.7% in 2018 (KISDI, 2018). This implies that while the major recipient countries of ICT ODA were located in Asia in the past, ICT ODA now has a diversified portfolio to include other various regions.

Korea is already acknowledged as the leading country in e-government, as well as being the top provider of ICT ODA to recipient countries with nearly 55% of the OECD DAC's aid in the ICT sector according to creditor reporting system (CRS). Korea is evaluated as having a comparative advantage, particularly the in ICT field such as e-government and e-learning. These sectoral priorities have already taken up large segments

Table 1 ICT ODA Budget by sector (One hundred million KRW, KISDI 2018)

	2015	*2016*	*2017*	*2018*	*Total*
ICT ODA	1,735	1,183	2,297	2,695	7,910
Pure ICT ODA'	836 (48.2%)	836 (48.2%)	534 (23.3%)	630 (23.4%)	2,429 (30.7%)
ICT & Public Administration	598 (34.5%)	435 (36.7%)	878 (38.3%)	631 (23.4%)	2,542 (32.1%)
ICT & Education	92 (5.3%)	93 (7.8%)	185 (8.1%)	413 (15.3%)	783 (9.9%)
ICT & Environment and Energy	77 (4.4%)	69 (5.8%)	97 (4.2%)	157 (5.8%)	400 (5.1%)
ICT & Economic and Industrial Infrastructure	66 (3.8%)	122 (10.3%)	330 (14.3%)	788 (29.2%)	1,306 (16.5%)
ICT & Agriculture, Forestry, and Fisheries	14 (0.8%)	6 (0.5%)	85 (3.7%)	30 (1.1%)	135 (1.7%)
ICT & Others (Health, etc.)	52 (3.0%)	30 (2.5%)	187 (8.1%)	46 (1.7%)	315 (4.0%)

Table 2 ICT ODA Budget by region (One hundred million KRW, KISDI 2018)

	2015	*2016*	*2017*	*2018*	*Total*
ICT ODA	1,735	1,183	2,297	2,695	7,910
Asia	919 (53.0%)	474 (40.0%)	714 (31.1%)	871 (32.3%)	2,978 (37.6%)
Africa	335 (19.3%)	309 (26.1%)	871 (37.9%)	620 (23.0%)	2,135 (30.0%)
Middle East/CIS	196 (6.1%)	126 (10.7%)	430 (18.7)	612 (22.7%)	1,364 (17.2%)
Latin America	126 (7.3%)	119 (10.1%)	114 (5.0%)	364 (13.5%)	723 (9.1%)
Others	249 (14.4%)	155 (13.1%)	168 (7.3%)	228 (8.5%)	800 (10.1%)

of Korea's ODA carried out by EDCF loans and Korea International Cooperation Agency (KOICA) grants. However, there is still a missing empirical link between prioritization of ICT and better outcomes in local context. Schopf (2019) argues that Korea's ICT ODA lacked governance, transparency, clear goals, quantifiable measurement, and independent evaluation. Still, Korea's recent prioritization of ICT has brought international attention. Enhancing good governance and implementing ICT innovations followed by a result-based framework will make Korea further realize its potential to utilize ICT for development effectiveness.

3.3 Implementation of K-SDGs and STI

The K-SDGs were implemented in 2018. The 17 goals and 122 targets of the K-SDGs, which aim to effectively implement the SDGs within the domestic politico-economic circumstances of Korea, were set under the supervision of the Commission on Sustainable Development in Korea. Given the cross-cutting nature of STI, a wide range of potential roles that STI can play has been recognized in government. A couple of specific SDGs are directly related to STI such as health (SDG3), education (SDG4), water (SDG6), clean energy (SDG7), work and economic growth (SDG8), industry, innovation and infrastructure (SDG9), sustainable cities (SDG11), responsible consumption and production (SDG12), and climate change (SDG13). STI is also incorporated as leverage for building capacity and knowledge sharing in other goals including peace, justice, and institutions (SDG16); and global partnerships (SDG17).

In Korea, the Ministry of Science and ICT (MSIT) is the leading government body of R&D and innovation policy. In the implementation process of the K-SDGs, the MSIT participated in the Consultative Committee of the Related Ministries, but did not play a vital role in shaping targets and indicators of the K-SDGs. Only the target of the improvement of national STI competitiveness by increasing STI resources (SDG9.4) was designated to the MSIT. In addition, the MSIT has been involved other targets such as 'promote mental health and drug abuse and misuse' (SDG3.2) with the Ministry of Health and Welfare; 'accessibility to sustainable infrastructure' (SDG9.1) with the Ministry of Land, Infrastructure, and Transport (MOLIT) and the Statistics Korea; 'establish and carry out policies for sustainable consumption and production' (SDG12.1) with the Ministry of Economy and Finance (MOEF), the Ministry of Environment, the Ministry of Trade, Industry, and Energy, and the Ministry of SMEs and Startups, and finally 'support developing countries to strengthen system for science and technology innovation' (SDG17.3) with the MoFA, the MOEF, and the Commission on Sustainable Development. This type of complex responsibility has been a criticism of fragmentation in Korea's development strategy which will be discussed in the next section.

In terms of budget and strategy, STI has not been a major concern in Korean ODA policy compared to other main donor countries. As explained above, the assigned engagement of MSIT in SDGs from the K-SDG framework looks quite limited, despite high expectations for the STI role in contributing to achieving the SDGs. Furthermore, the ODA budget in MSIT has been marginal in Korea, whereas the growing importance of STI in the SDGs has been more recognized in recent years. From an institutional view, the MSIT may not be the major body of the K-SDGs nor Korea's ODA policy. However, it is generally understood that the MSIT is the core of national STI and development policy because of its traditional role of promoting science and technological capabilities in the process of national economic development.

3.4 *Toward Integration in STI-related ODA*

There has been criticism of fragmentation between the Ministry of Foreign Affairs (MOFA), which has driven 'science for diplomacy' agenda and Ministry of Science and ICT (MSIT), which has led science and technology international cooperation. MOFA has tended to use science and technology as a foreign policy tool and a soft power, whereas MSIT has regarded the cooperation with developing countries as a minor policy priority compared to R&D and high-tech cooperation with advanced countries and industrial-university cooperation (Chang, 2012).

In this regard, the MSIT and the MOFA began discussions in October 2018 and announced the 'National Science and Technology Diplomacy Strategy for Innovative and Inclusive State' on October 30, 2019, which highlights the role of STI-related ODA and international cooperation in achieving the SDGs. This plan was a joint strategy established by MSIT, the STI-driven ministry, and MOFA, the ODA-managing ministry. This plan sets three main goals to co-create a future for humanity, promote global shared growth, and lastly pursue safeguarding citizens' everyday life (Ministry of Science and ICT, 2019). Those goals are to be pursued with four main strategies, (a) leading the global agenda and promoting national interests; (b) contributing to sustainable development in the international community; (c) improving national security and the quality of life of the people; and (d) institutionalizing the governance structure.

In order to implement these strategies, the government aimed to (i) gradually establish a science and technology diplomacy support system that embraces science and technology expertise and diplomatic networks; (ii) expand cooperation with other countries; (iii) support ODA by utilizing science and technology for global shared growth; (iv) enhance cyber security and disaster response system of overseas consulates and government agencies; and (v) enhance ministerial coordination to solve the fragmentation issue. The introduction of this vision was relatively well constructed, yet the tangible changes in policy and achievements were shown to be insignificant.

Due to the cross-cutting nature of STI-ODA, policy and project level integration between key supervisory ministries and executing governmental agencies for ODA has been encouraged in Korea. There are roughly four different cases of project integration in ODA: (i) sectorally similar projects joined together with the neighboring regions; (ii) diverse projects to be converged in the same region; (iii) sequential approach to regional integration of individual projects; and (iv) individual projects redesinged as one integrated program (Kim et al.,2019). Kim et al. (2019) provide examples including the agriculture sector where agricultural technology education and consulting are adopted through the cooperation with the Ministry of Agriculture and Forestry, KOICA, and Ministry of the Interior and Safety (MoIS), and the cultural sector with heritage recovery, tourism, and education projects through the collaboration with the MoIS, Cultural Heritage Administration and Ministry of Culture, Sports and Tourism. However, limitations on the integration of different governmental departments still exist. Unless there are specific implementation plans and well-arranged guidelines from the early stages, the issues will remain unsolved.

In addition to institutional building toward a coordinated mechanism for STI in ODA policy, there have been trials and efforts at the project level. Among project-based STI-ODA initiatives, KOICA's Creative Technology Solution (CTS) program—the main program of 'Inclusive Innovation'—is the most representative. The initiative targets the bottom of the pyramid (BOP) in developing countries, generating sustainable business models using creative and innovative technologies, and also launching exemplary businesses through social entrepreneurship. Since 2015, 76 projects were implemented by CTS, and at BOP in developing countries, 1.3 million people were direct and indirect beneficiaries (KOICA, 2021). KOICA (2019) has supported a total of 23 projects in 20 countries in 2018; 17 in Asia, five in Africa, and one in the Middle East. Due to the government's emphasis on New Southern Policy, Asia has been the main specific region in which to conduct the projects (KOICA, 2019). Although rigorous assessment considering the local context is needed, there are a couple of notable cases. According to the case of a mobile-based malaria diagnostic kit, which was selected as one of the CTS programs, the project is expected to expand an innovative way for start-ups and also devote its technology in achieving the SDG in developing countries (Lee & Yim, 2016).

Based on the positive internal evaluation of these initiatives in the agency, a longer-term plan has been suggested that scales up national-level businesses by connecting social and economic development initiatives. CTS is meaningful as an innovative and explorative approach for STI-ODA at the project level. However, impact evaluations on individual projects are still lacking, and there is little empirical evidence for better development effectiveness. A wider range of disruptive technologies such as artificial intelligence, big data, satellite imagery, remote sensing, drone technologies, and robotics have been extensively tested and applied to development projects by various development agencies (Park, 2020; Park et al., 2020). One of the key factors for success is the localization of technology considering contextual differences in developing countries. Also, it is crucial to emphasize the need for supporting commercialization of innovative technologies and encourage the active participation of local actors in the implementation process of technologies.

In STI-ODA, one of the most important stakeholders is the private sector. At the 2011 High Level Forum on Aid Effectiveness in Busan, the public-private partnership was highlighted as an important agenda in the Busan Partnership document. Various stakeholders' engagement suggested a new way to expand partnerships in development cooperation, share financial and human resources, and increase development effectiveness through risk sharing. KOICA and the Export-Import Bank of Korea (EXIM Bank) also sought active participation from private sectors and social entrepreneurs through impact investment and corporate social responsibility (CSR). However, since the projects proposed in the public sectors were based on government-led contracts, some cases were subjected to criticism for being considerably time-consuming as well as losing agility and efficiency.

4 Emerging Agenda in STI-driven Development Cooperation

4.1 *Tackling the Pandemic with STI*

The COVID-19 outbreak throughout the world created an unprecedented health crisis globally and nationally. The COVID-19 pandemic has accelerated the shift of focus of ODA on STI. While the implementation of the SDGs is a complex process, the global pandemic has added another layer of complexity and is threatening the international and national development efforts. The pandemic has plunged global health, development, and humanitarian organizations into a prolonged and worldwide crisis, and has begun to displace other global priorities. The World Bank/IMF Development Committee (2020) outlined in their communique the devastating impacts of COVID-19 on the global economy with disruptions to trade, supply chains, and investment flows as well as impacts on human well-being, income loss for households, and disruptions in the delivery of essential public services. COVID-19 has the potential to erase the existing development gains for many countries (IISD, 2020; World Bank/IMF Development Committee, 2020). In response to the COVID-19 pandemic, more than 50 countries have announced and implemented some forms of disaster relief allowances and social assistance to tackle the immediate challenges faced by their citizens (Gelb & Mukherjee, 2020). Governments have also explored how the application of digital technologies can help accelerate global efforts to achieve the SDGs.

Countermeasures for COVID-19 have revealed the government's crisis response capabilities, resilience, and flexibility to adapt during the pandemic. South Korea was relatively successful in controlling and mitigating the pandemic without the need for a lockdown. The Korean government opted for unique agile-adaptive and transparent actions, as well as multi-collaborative governance to mitigate the surge of the pandemic (Moon, 2020). The hyper-network environment among the government, businesses, and citizens led to Korea's success in 'flattening the curve' by following the 3T policy (test, trace, and treat) (Heo et al., 2020). The Korean government quickly shifted its focus to tracing and quarantine measures to contain the pandemic. The Ministry of Land, Infrastructure, and Transport (MOLIT) began developing the COVID-19 Smart Management System (SMS) through smart city hub technology in mid-February with the Korean Center for Disease Control (KCDC) and the MSIT (MSIT, 2020). The system collects big data, such as mobile phone location, hot spot usage, CCTV recordings, and credit card usage within 10 minutes (Ministry of Land and Infrastructure, 2020). As a measure to protect personal information, the Epidemic Investigation Support System (EISS) has limited access to epidemiology investigators to identity routes of only positively tested cases (Park, 2020; Park et al., 2020).

As discussed above, in Korea, there has been a relative lack of attention to STI in ODA, in particular emerging digital technologies in ODA programs. The COVID-19 crisis revealed how digital technologies could be leveraged in managing the COVID-19 pandemic and promoting digital resilience in developing countries (Park et al., 2021). The MoFA shared Korea's experiences and lessons learned from the COVID-19 crisis by highlighting the application of data sharing platforms and artificial intelligence in response to the pandemic at the OECD DAC high-level meeting in November 2020. The development of policies and capacities for applying emerging technologies has also become a priority in development cooperation strategy in Korea. To reflect this shift and achieve new objectives in ODA policy, the government has directed focus to digital transformation and more proactive engagement from the private sector. Investments in R&D, particularly in the ICT and energy sector and public-private partnerships will be aimed at leveraging the fourth industrial revolution in order to create better opportunities in development cooperation in the international arena and to foster inclusive and sustainable growth in the domestic arena. Recently, Korea introduced 'the Fourth Industrial Revolution for inclusive society' as part of the presidential agenda, while Japan introduced 'Future Vision Towards the 2030s' subtitled as the vision for a new industrial structure. This change has accelerated the implementation of STI, in particular, emerging digital technologies in Korean ODA. Policy for emerging technologies is a primary policy directive of the Korean government coordinated through a presidential committee such as the Committee of the Fourth Industrial Revolution, and the National Council of Science and Technology. The uniqueness of Korean STI policy includes large investments in R&D and incentives for attracting private sector investments in sector-specific areas. The scope and aims of research in science and technological innovation have been gradually broadened and greatly influenced the landscape of Korea's ODA policy.

4.2 Highlighting STI in the Strategic Plan for International Development Cooperation

These kinds of discussions regarding STI–ODA were reflected in the establishment of the Strategic Plan for International Development Cooperation in 2020. The Strategic Plan for International Development Cooperation clarifies the basic framework for Korea's ODA policy (i) to take responsibility as a member of the OECD DAC (DAC), (ii) to meet its commitment of scaling up the ODA volume, and (iii) to strengthen integrated ODA governance. The third strategic plan (2021–2025) was approved by the National Committee for International Development Cooperation in 2020. Compared to the previous plans, the third strategic plan highly emphasizes STI according to four main ideas. First, given the situation of the COVID-19 pandemic, Korea comprehensively stresses a plan to support the provision of healthcare and multifaceted infectious disease response plans in recipient countries and will disseminate essential drugs and technologies related to test, trace, and treat (3T). Second, in 2020, the Korean government announced the Digital New Deal based on the importance of contact-free industries and digital transformation. Based on this, some opinions were suggested to introduce ICT in development cooperation, including the aforementioned health sector. In this way, the Korean government intends to foster ODA linked to the Digital New Deal to establish an ICT infrastructure by supporting an open data system and e-government in developing countries, in line with the digital transformation of the recipient country undergoing change due to the rise of a contact-free economy and the expansion of culture. Third, considering the demand from recipient countries in energy-related industries, the government also encourages the transition to a low-carbon economy by focusing on green technology such as smart farms and renewable energy, which could eventually be linked to the Green New Deal. Lastly, enhancing developing countries' overall science and technology human resources, developing infrastructure research capacity and policy, and addressing the countries' digital divide issues were also included as important agenda points.

4.3 *Aligning Korea's Strategy to the Global Framework*

It is crucial for the Korean government as an emerging donor to actively participate in the international debate on STI-driven development cooperation. As discussed in Sect. 2.4, a mechanism for incorporating STI components in development initiatives has been developed in the global field of development. Based on the mandate discussed in paragraph 124 of AAAA, the TFM as an implementing structure was established. In addition, according to the STI Forum, the initiative to develop STI for an SDG roadmap is being driven by the UN Interagency Task Team on Science, Technology, and Innovation for the SDGs (IATT). Members include key UN agencies such as UN-DESA, UNCTAD, UNESCO, UNWIPO, UNIDO, UN-ESCAP, and UNU as well as the World Bank. Substantial financial and political support has been offered by international stakeholders, including the Global Sustainable Technology and Innovation Conference Series, the European Commission's Joint Research Centre (JRC), the OECD, and Japan. The African Union Commission has endorsed the IATT and expressed interest in implementing the roadmaps in African countries.

Building on these global efforts, at the national level, national guidelines on STI for the SDGs have been suggested. It is key for a national government to implement a globally agreed framework for the SDGs as well as further develop and internalize its own national STI agenda. With this backdrop, it is also important to utilize this framework and international guidelines in the implementation of Korea's STI–ODA in developing countries. From these efforts of the global community, the IATT has developed the 'Guidebook for the Preparation of Science, Technology and Innovation (STI) for SDGs Roadmaps' (IATT, 2020). The guidebook outlines the recommended guidelines for developing national STI for SDG roadmaps according to the six key steps described in Fig. 1, as well as reports for five countries with different SDG gaps and STI capabilities. There are currently undergoing pilot programs for the STI for SDGs roadmaps initiative in Ethiopia, Ghana, India, Kenya, and Serbia.

Furthermore, the IATT guidebook also outlines recommendations for the best approaches to STI road-mapping and recommends that the process be done at three main levels, from the subnational to the national and then to the international level following the key steps outlined in Fig. 2. With better coordination across the different levels constituting a collective learning policy, these institutional frameworks can provide

Fig. 1 Newly proposed process flow of 6 key steps in the STI for SDGs roadmaps (IATT, 2020)

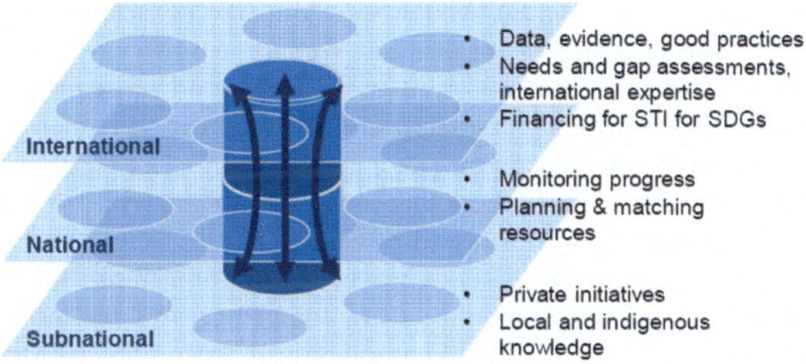

Fig. 2 Bridging the different institutional level for STI roadmap (IATT, 2020)

policy implications and enhance alignment between Korea's development programs and local governance.

4.4 *Future Policy and Research Agenda*

As discussed in Sect. 3, policy dialogue on the importance of STI in Korea's ODA began with the time of joining the OECD DAC in 2010 and hosting the High Level Forum on Aid Effectiveness in Busan in 2011. Since then, the proliferation of STI, particularly ICT-related development projects, has given rise to new opportunities and challenges in development cooperation. However, there have been very few scholarly works on STI-ODA in the context of Korea. Kim et al. (2019) note that domestically there are only 59 research articles on STI-related ODA as of 2019. Among them, there are very few peer-reviewed academic journal articles. They attribute this to research projects being small-scale, as intermittent international development projects of government-funded research institutes in the Science and Technology sector. However, there are three general reasons for the lack of research on STI and development cooperation in the context of Korea. First, as it is in a global context, STI-related ODA statistics are difficult to produce with STI being a cross-cutting issue. Second, institutions carrying out STI-related ODA are very much fragmented in Korea, including not only Korea International Cooperation Agency (KOICA), EXIM-BANK, Ministry of Science and ICT (MSIT), but also Ministry of Health and Welfare, Ministry of Interior (e-government), and other lines of ministries. Third, and most importantly, a systematic understanding, research, and analysis on STI-related ODA has not been established.

Despite such challenges, there has been policy analysis on STI and international cooperation in a broad sense. Especially, Science and Technology Policy Institute (STEPI), a government think-tank, has consistently conducted policy projects relating to science and technology for diplomacy and international R&D cooperation since the 2000s. Also, Korea Institute for International Economic Policy (KIEP) and Korean National Diplomatic Academy among others have conducted related research (Chang, 2012). Discussion on STI-ODA has become more vibrant ever since 2010 when Korea joined the OECD DAC. STEPI's initial set of policy research included case studies on ASEAN nations (Lee et al., 2010), the establishment of strategic analysis on the Korean STI-ODA Roadmap (Lee et al., 2012; Lee, 2011; Yim, 2015). While the stream of research has focused on the establishment of a 'Korean ODA model' strategy, the limitations were that they were confined to a technology-centric perspective and its linear application to developing countries.

Moving from findings from Korea's case and the discussion of opportunities and challenges of STI in development cooperation, this study identifies major future policy and research agenda as below. First, developing measurement and indicators of STI-ODA is urgent. Second, agile policy and exploratory academic research on the impact of digital transformation and datafication will be increasingly important. Third, empirical evidence and impact evaluation on innovative solutions of technology such as KOICA's CTS program are needed. Fourth, there is also a need for contextualizing STI-ODA for local embeddedness and integrating the strategic initiatives for better development management.

5 Conclusion

STI policies have played an irreplaceable role in socioeconomic transformation and industrialization. Korea has also progressed through different phases of socioeconomic and industrial development. During the industrialization period, the government of Korea enhanced domestic technological capacity by promoting efficient technology policies while facilitating foreign direct investment and technology assistance from developed countries. With the backdrop of national development in Korea demonstrating the significance of endogenous technological innovation, STI became increasingly important in Korea's ODA policy as well. In 2010, Korea joined the OECD DAC and prioritized ICT components in its ODA strategies. The increasingly rapid pace of technological change and Korea's global competitiveness in the ICT sector have driven the reorientation of ODA programs to promote an ICT-focused Korean ODA model.

However, compared to other major donor countries, STI has played a minimal role both financially and strategically in Korean ODA policy. The ODA activities related to STI have also sporadically followed policy changes in terms of the focus of aid from e-government, ICT, and health sectors in a fragmented manner. As Korea's response to COVID-19 in 2020 received global attention, especially from developing countries, the importance of digital and data-driven public health cooperation is in the spotlight. Based on the Green New Deal and Digital New Deal initiative launched in June 2020, it is important to note that the government is currently structuring a three-pronged STI–ODA strategy based on the themes of digitization, green technology, and health technology. The Korean government announced in early 2021 that it will expand the share of ODA to climate and digital transformation from the current 6.4% to 22.7% of its total ODA as part of South Korea's Global Strategy on the Green and Digital New Deal. Building on the lessons from the case of Korea, four main policy and research implications are derived.

First, in spite of the recognition of the importance of STI in ODA, there is still a lack of clear understanding of STI in the national strategy for development cooperation. In the absence of a strategic approach to embracing STI in national ODA strategies, it is unlikely that Korea will be able to benefit from the potential of STI. STI-driven ODA programs need a comprehensive approach including financial resources, investment, and technology governance to apply such STI to the very complex local context.

Second, STI actors and agencies need to integrate a demand-driven perspective. The notion of technology transfer overemphasizes a technical and functional process, but often ignore socio-organizational context in developing countries. STI policy and strategies in developing countries are shaped in a very complex way that involves contextualization, incentive restructuring, and shifting power dynamics (Avgerou & Walsham, 2000; Orlikowski, 1992). The implementation process of technology is subject to internal and external events involving diverse local and international actors and their interactions. Thus, it is important to understand that the alignment of particular STI–ODA to local interests cannot be simply pre-planned and controlled following rational economic assumptions, but is constantly constructed through improvisations, political negotiations, bottom-up demand, and sometimes external shocks such as the global pandemic.

Third, in terms of the measurement and evaluation of STI–ODA, the cross-cutting nature of STI makes it an important factor in achieving nearly all the SDGs. Measurement of total STI development, however, remains a challenge. Ericsson & Mealy (2019) present an attempt to quantify STI efforts by estimating the financial resources dedicated to STI development activities through the creditor reporting system. Also, there have been attempts to set standards for measuring STI-ODA projects in Korea by Korean national research institutes (Kang & Yim, 2014; Kim et al., 2019). Significant differences in definitions and classification between data sources cause both underestimates and overestimates of the total financial resources spent on STI. Ericsson & Mealy (2019) highlight the need for the STI sector and development communities to start a dialogue to align their definitions and values toward effective STI development monitoring. Improvement of the CRS sector code is suggested as a possible solution, together with the use of machine learning and/or adopting other statistical methods.

Fourth, as discussed, science, technology, and innovation are overlapping concepts with strong interlinkages, making them crucial elements in achieving the SDGs. Many SDGs have objectives that are directly related to STI, while others rely on the development of knowledge and technology and are therefore indirectly connected to STI. As the AAAA calls for financial commitment and increased ODA and emphasizes public–private partnerships to create national STI strategies, investment in STI should not be regarded as a target itself, but more as a means to achieve the SDGs and ensure sustainable development in developing countries. In particular, increased investments in digital transformation, including in infrastructure and skills, as well as cooperation for the revision of regulations that impede technological innovation and new business models in developing countries, are crucial.

There is a shortage of empirical research on STI–ODA in Korea, with the discussion limited to policy research and strategic initiatives toward the Korean ODA model. This study provides a foundational discussion on Korea's future STI–ODA policy. Also, practices in Korea show how STI may create common interests for both developed and developing countries through multi-layered cooperation as well as delivering practical contributions to solving global challenges. Building on the lessons and challenges in the case of Korea, implications for development policy may include (1) improving policy integration and interlinkages between ODA and STI policies, including in financing and governance structure; (2) adopting STI for the localized applications that meet the country's specific socioeconomic conditions and priorities; (3) developing national STI roadmaps and measurement frameworks; and (4) balancing the import of innovative solutions and local development, as well as encouraging private sector engagement.

References

Avgerou, C. (2003). The link between ICT and economic growth in the discourse of development. In K. Mikko, M. Ramiro, & P. Angeliki (Eds.), *Organizational information systems in the context of globalization* (pp. 373–386). Springer.

Avgerou, C. (2010). Discourses on ICT and development. *Information technologies & international development, 6*, 1–18.

Avgerou, C., & Walsham, G. (Eds.). (2000). *Information technology in context: Implementing systems in the developing world*. Ashgate.

Baark, E., & Heeks, R. (1999). Donor-funded information technology transfer projects: Evaluating the life-cycle approach in four Chinese science and technology projects. *Information Technology for Development, 8*(4), 185–197. https://doi.org/10.1080/02681102.1999.9525309

Chang, Y. S. (2012). Packaged science and technology ODA models as new-science and technology diplomacy strategies. *STEPI Insight*, 86, 1–29.

Correa, C. (1994). Trends in technology transfer implications for developing countries. *Science and Public Policy, 21*(6), 369–380.

Ericsson, F., & Mealy, S. (2019). *Connecting official development assistance and science technology and innovation for inclusive development: Measurement challenges from a development assistance committee perspective*.

Forsyth, T. (2007). Promoting the "development dividend" of climate technology transfer: Can cross-sector partnerships help? *World Development, 35*(10), 1684–1698. https://doi.org/10.1016/j.worlddev.2007.06.001

Fukuda-parr, S., & Muchhala, B. (2019). The Southern origins of sustainable development goals: Ideas, actors, aspirations. *World Development, 126*(November), 104706. https://doi.org/10.1016/j.worlddev.2019.104706

Furman, J. L., Porter, M. E., & Stern, S. (2002). The determinants of national innovative capacity. *Research Policy, 31*(6), 899–933. https://doi.org/10.1016/S0048-7333(01)00152-4

Gamser, M. S. (1988). Innovation, technical assistance, and development: The importance of technology users. *World Development, 16*(6), 711–721. https://doi.org/10.1016/0305-750X(88)90177-5

Gelb, A., & Mukherjee, A. (2020). *Can we use digital technology to cushion the pandemic's Blow—And in the longer run, deliver on the SDGs?* Center For Global Development.

Gibson, C. C., Hoffman, B. D., & Jablonski, R. S. (2015). Did aid promote democracy in Africa? The role of technical assistance in Africa's transitions. *World Development, 68*, 323–335. https://doi.org/10.1016/j.worlddev.2014.11.009

Glass, A. J., & Saggi, K. (1998). International technology transfer and the technology gap. *Journal of Development Economics, 55*(2), 369–398. https://doi.org/10.1016/S0304-3878(98)00041-8

Godfrey, M., Sophal, C., Kato, T., Vou Piseth, L., Dorina, P., Saravy, T., & Sovannarith, S. (2002). Technical assistance and capacity development in an aid-dependent economy: The experience of Cambodia. *World Development, 30*(3), 355–373. https://doi.org/10.1016/S0305-750X(01)00121-8

Guston, D. H. (1999). Stabilizing the boundary between US politics and science: Teh role of the office of technology transfer as a boundary organization. *Social Studies of Science, 29*(1), 87–111.

Hahm, S. D., Plein, L. C., & Florida, R. (1994). The politics of international technology transfer: Lessons from the Korean experience. *Policy Studies Journal, 22*(2), 311–321. https://doi.org/10.1111/j.1541-0072.1994.tb01470.x

Heeks, R. (2008). ICT4D 2.0: The next phase of applying ICT for international development. *Computer, 41.* https://doi.org/10.1109/MC.2008.192

Heo, K., Lee, D., Seo, Y., & Choi, H. (2020). Searching for digital technologies in containment and mitigation strategies: Experience from South Korea Covid-19. *Annals of Global Health, 86*(1), 1–10. https://doi.org/10.5334/AOGH.2993

IATT. (2020). *Guidebook for the preparation of science, technology and innovation (STI) for SDGs roadmaps.*

IISD. (2020). Virtual world bank, IMF meetings recognize COVID-19's impact on development | News | SDG Knowledge Hub | IISD.

Juma, C., Fang, K., Honca, D., Perez, J. H., Konde, V., Lee, S. H., Arenas, J., Ivinson, A., Robinson, H., & Singh, S. (2005). Global governance of technology: Meeting the needs of developing countries. *International Journal of Technology Management, 22*(7/8), 629. https://doi.org/10.1504/ijtm.2001.002982

Kang, H. J., & Yim, D. S. (2014). Current status and policy direction of Science and Technology, & ICT ODA. *STEPI Insight, 145*, 1–37.

Kapur, D. (2001). Diasporas and technology transfer. *Journal of Human Development, 2*(2), 265–286. https://doi.org/10.1080/14649880120067284

Karakosta, C., Doukas, H., & Psarras, J. (2010). Technology transfer through climate change: Setting a sustainable energy pattern. *Renewable and Sustainable Energy Reviews, 14*(6), 1546–1557. https://doi.org/10.1016/j.rser.2010.02.001

Kim, W., Lee, J. W., Yim, D. S., Kim, K. K., Sun, I., Gang, H. J., Kim, E., Kwon, S., Jeon, B. W., & Lee, S. (2019). *Analysis and strategies on science and technology ODA.*

KISDI. (2018). *ICT ODA evaluation and monitoring.*

Korea Institue for International Economic Policy. (2015). *Research on establishing korean science and technology ODA roadmap.*

Korea International Cooperation Agency. (2019). *2019 KOICA creative technology solutions brochure.*

Korea International Cooperation Agency. (2021). *2021 KOICA CTS programme guide book.*

Lee, D. Y., & Yim, C. Y. (2016). Case study on Korea International Cooperation Agency Creative Technology Solution (KOICA CTS) Programme: Next-generation mobile malaria diagnosis kit. *Journal of International Development Cooperation, 4*, 41–52.

Lee, J. hyop. (2011). Korea science, technology and innovation ODA strategies for global platform leadership. *Science & Technology Policy, 185*, 24–36.

Lee, J. hyop, Dong, G., Yang, F., Sun, J., & Maliphol, S. (2012). *Korean Science, Technology and Innovation (STI) ODA Strategies.*

Lee, W. S., Hwang, J. T., & Kim, W. J. (2010). *A framework to analyse and enhance S&T ODA for sustainable development in three ASEAN countries: Cambodia, Vietnam, and Philippines.*

Ministry of Education, Science and Technology. (2011). The national strategy of science, technology, and education ODA for enhancing developing countries' capacity and supporting sustainable development.

Ministry of Land and Infrastructure. (2020, March 25). Fast and precise detection of COVID-19 infected patient travel history with smart city technology. *Ministry of Land and Infrastructure Press Release.*

Ministry of Science and ICT. (2019). Korean government announced diplomatic strategies on science and technology for the innovative & inclusive state. Republic of Korea Policy Briefing.

Ministry of Science and ICT. (2020). *How we fought COVID-19: Perspective from science & ICT*. Sejong.

Miedzinski, M., Mcdowall, W., Fahnestock, J., Muller, G., & Lopez, F. J. D. (2019, November). *Science, technology and innovation policy roadmaps for the SDGs a guide for design and implementation.*

Moon, M. J. (2020). Fighting COVID-19 with agility, transparency, and participation: Wicked policy problems and new governance challenges. *Public Administration Review, 80*(4), 651–656. https://doi.org/10.1111/puar.13214

Murphy, K., Kirkman, G. A., Seres, S., & Haites, E. (2015). Technology transfer in the CDM: An updated analysis. *Climate Policy, 15*(1), 127–145. https://doi.org/10.1080/14693062.2013.812719

Nam, D. W. (1997). *Korea's economic growth in a changing world*. Samsung Economic Research Institute.

Nelson, R. (1993). *National systems of innovation: A comparative study*. Oxford University Press.

Oda, H. (1991). *Law and politics of West-East technology transfer* (H. Oda, Ed.). Graham & Trotman. Martinus Nijhoff.

OECD. (1996). *The Knowledge-based economy.*

OECD/Eurostat. (2018). *Oslo manual 2018: Guidelines for collecting, reporting and using data on innovation* (4th ed.). Luxembourg: OECD Publishing. https://doi.org/10.1787/9789264304604-en

Orlikowski, W. J. (1992). The duality of technology: Rethinking the concept of technology in organizations. *Organization Science, 3*(3), 398–427.

Park, K. R. (2020). Data revolution: A critical view from the perspective of development cooperation. *International Development and Cooperation Review, 12*(2), 1–20.

Park, K. R., Sahay, S., Braa, J., & Amarakoon, P. (2021). *Digital resilience for what? Case study of South Korea.* In International Conference on Information and Digital Technology for Development: ICT4D. Springer.

Park, Y. J., Cho, S. Y., Lee, J., Lee, I., Park, W. H., Jeong, S., Kim, S., Lee, S., Kim, J., & Park, O. (2020). Development and utilization of a rapid and accurate epidemic investigation support system for Covid-19. *Osong Public Health and Research Perspectives, 11*(3), 118–127. https://doi.org/10.24171/j.phrp.2020.11.3.06

Reddy, N. M., & Zhao, L. (1990). International technology transfer: A review. *Research Policy, 19*(4), 285–307. https://doi.org/10.1016/0048-7333(90)90015-X

Romer, P. M. (1990). Endogenous technological change. *Journal of Political Economy.* https://doi.org/10.3386/w3210

Rostow, W. W. (1960). *The stages of economic growth: A non-communist manifesto.* Cambridge University Press.

Sahay, S., Sein, M., & Urquhart, C. (2017). Flipping the context: ICT4D, the next grand challenge for IS research and practice. *Journal of the Association for Information Systems, 18*(12).

Schopf, J. C. (2019). Room for improvement: Why Korea's leading ICT ODA program has failed to combat corruption. *Telecommunications Policy, 43*(6), 501–519. https://doi.org/10.1016/j.telpol.2019.01.001

Schot, J., & Steinmueller, W. E. (2018). Three frames for innovation policy: R&D, systems of innovation and transformative change. *Research Policy, 47*(9), 1554–1567. https://doi.org/10.1016/j.respol.2018.08.011

Sharif, N. (2006). Emergence and development of the national innovation systems concept. *Research Policy, 35*(5), 745–766. https://doi.org/10.1016/j.respol.2006.04.001

Soete, L. (2019). Science, technology and innovation studies at a crossroad: SPRU as case study. *Research Policy*, *48*(4), 849–857. https://doi.org/10.1016/j.respol.2018.10.029

Solow, R. M. (1956). A contribution to the theory of economic growth. *The Quarterly Journal of Economics, 70*(1), 65–94. https://doi.org/10.2307/1884513

Tarnoff, C. (2018). *The Marshall plan: Design, accomplishments, and significance.*

United Nations. (2018). *Technology and the UN 2030 agenda: United Nations Technology facilitation mechanism.*

United Nations. (2015). *Addis Ababa Action Agenda of the third international conference on financing for development (Addis Ababa Action Agenda).* A/RES/69/313.

Walsh, P. P., Murphy, E., & Horan, D. (2020). The role of science, technology and innovation in the UN 2030 agenda. *Technological Forecasting and Social Change*, *154*(February 2019), 119957. https://doi.org/10.1016/j.techfore.2020.119957

Wilkins, G. (2002). *Technology transfer for renewable energy: Overcoming barriers in developing countries*. Earthscan, London.

Wilson, G. (2007). Knowledge, innovation and re-inventing technical assistance for development. *Progress in Development Studies, 3*(3), 183–199.

Wittemyer, R., Bailur, S., Anand, N., Park, K. R., & Gigler, S. (2014). New routes to governance: A review of cases in participation, transparency, and accountability. In *Closing the feedback loop: Can technology bridge the accountability gap?* World Bank.

World Bank. (2011). *Innovation in disaster risk financing for developing countries: Public and private contributions*. The World Bank.

World Bank/IMF Development Committee. (2020). *World Bank/IMF Spring Meetings 2020: Development Committee Communiqué.*

Yim, D. S. (2015). Korea science, technology and innovation ODA strategies and action plans. *Journal of International Development Cooperation, 2015*(3), 21–37.

Yoo, J. S., & Yoo, S. H. (2019). Diagnosis on Korea ICT ODA: Analysis on recent 5 Years of comprehensive implementation plan on international development cooperation (by fixed amount). *Information and Communications Broadcasting Policy, 31*(5), 59–121.

PART III

Emerging Agendas and New Challenges for International Development

CHAPTER 9

Civil Society Organizations as Partners and Critics of Japan's Aid Policy

Akio Takayanagi

1 Introduction

Civil society organizations (CSOs) or Non-governmental organizations (NGOs)[1] are important actors in international development. There have been debates on what it means by civil society or CSOs, what kind of groups should be included and what should be not, etc.[2] Also, there

[1] CSOs and NGOs are not synonymous. In the context of international development, CSOs include not only NGOs but also social movement groups, trade unions academic groups, faith-based groups, etc. part of whose work is international development. It is important that NGOs comprise only part of civil society, and expanding NGOs do not automatically mean strengthening civil society (Banks & Hulme, 2012: 21). It is since the beginning of the twenty-first century that the term CSOs was started to be widely used.

[2] For example, Hulme and Edwards eds. (1997); Van Rooy ed. (1998); Edwards (2011; 2020); Kaldor (2003).

A. Takayanagi (✉)
Department of Global and Intercultural Studies,
Ferris University, Yokohama, Japan
e-mail: takio@ferris.ac.jp

H.-j. Kwon et al. (eds.), *International Development Cooperation of Japan and South Korea*,
https://doi.org/10.1007/978-981-16-4601-0_9

have been debates between neo-liberal and radical understandings of civil society,[3] which I will not go into detail in this chapter.

The following definitions by OECD represent a general consensus on what it means by civil society and CSOs:

> Civil society refers to uncoerced human association and interaction by which individuals implement individual or collective action to address shared needs, ideas, interests, values and beliefs that they have identified in common, as well as the formal, semi-, or non-formal forms of associations and the individuals involved in them. Civil society is distinct from states, private for-profit enterprises, and the family.
>
> Civil society organisations (CSOs) are an organisational representation of civil society and include all not-for-individual profit, non-state, non-partisan, non-violent, and self-governing organizations outside of the family in which people come together to pursue shared needs, ideas, interests, and beliefs, including formal, legally registered organisations as well as informal associations without legal status but with a structure and activities. (OECD, 2021)

DAC members in Western Europe, North America, and Southern Pacific have partnered with NGOs/CSOs since the mid-1960s or 1970s. DAC members in total allocated about 15% of their bilateral official development assistance (ODA) in 2018. Nine member countries spent over 25% of ODA for CSOs, while the figure is only 1.7% for Japan and 2.1% for Korea (OECD, 2020a). Partnering with civil society is an area of weakness for the two East Asian DAC members, the focus of this book, although recent development in Korea was the launch of a policy framework on CSO-government aid agency partnership (KOICA, 2019; Kim and Hong this book).

In a recent OECD study on members' partnership with civil society, CSOs are acknowledged as independent development actors in their own right and as essential partners of official aid agencies in international development. DAC members have partnered with CSOs mainly for two objectives: to implement programs related to service delivery and to strengthen a pluralistic and independent civil society in partner

[3] See, for example, Kaldor (2003), Howell and Pearce (2001), Edwards (2011).

countries (OECD, 2020b: Chapter 2).[4] But it has always been questioned whether CSOs could fulfill their mission of social transformation while depending too much on official sources (Bebbington et al., 2008; Hulme & Edwards, eds. 1997; Banks and Hulme, 2012; Banks, Hulme & Edwards, 2013). Some analysts question whether CSOs, with official donors as one of the major funding sources, are independent—according to Tvedt (2006), NGOs are part of the "DOSTANGOS (Donors States and NGOs) system."[5]

It was only in 1989 that the Government of Japan (GoJ) established funding schemes for CSOs. This chapter examines the current issues around the partnership between CSOs and official aid agencies in Japan and identifies challenges for both CSOs and GoJ. Especially, I will examine whether the two objectives described in the OECD study are paid attention to in the partnership in Japan.

In identifying the challenges, reference will be made to especially the "action points" of the OECD study which were made "with a view towards developing new guidance or a recommendation to improve how members work with civil society sector (OECD, 2020b: 105)," and the "DAC Recommendation on Enabling Civil Society in Development Co-operation and Humanitarian Assistance" (hereafter the Recommendation) (OECD, 2021), adopted at the OECD-DAC on July 6, 2021. The Recommendation is the result of Paragraph 19 on the roles of CSOs of the communique of the OECD/DAC High Level Meeting (HLM) 2020 on 9–10 November which in the end says "we commit to developing a new DAC policy instrument[6] on enabling civil society (OECD, 2020d)."

[4] In early-2010s, there was another research report (OECD 2011) and a document which discusses lessons drawn from the DAC Peer Review regarding partnering with civil society (OECD, 2012). The twelve lessons in the document are; (1) Have an evidence-based, overarching civil society policy, (2) Strengthen civil society in developing countries, (3) Promote and support public awareness-raising, (4) Choose partners to meet objectives, (5) Make policy dialogue meaningful, (6) Respect independence while giving direction, (7) Match funding mechanisms with the purpose, (8) Minimize transaction costs, (9) Build strong partnerships with humanitarian NGOs, (10) Focus reporting on results and learning, (11) Increase transparency and accountability, and (12) Commission evaluations for learning and accountability.

[5] See Lewis, Kanji, and Themudo (2021) Chapter 8 for debates on CSOs-official donors' relations.

[6] The use of the word "instrument" in the DAC HLM communique was a result of compromise; DAC members could not reach agreement on whether it should be a

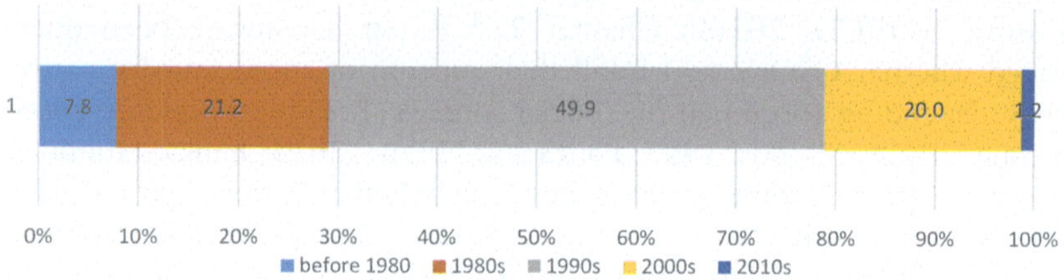

Fig. 1 When Japanese CSOs were founded (n = 345) (*Source* JANIC, 2016)

In Sections 2 and 3, I will quickly review the history of CSOs and their partnership with the government's aid agencies and the current trends of the CSO community. Section 4 will look into the GoJ's CSOs partnership policies and funding schemes for CSOs. Section 5 will be on policy dialogue between CSOs and GoJ. Section 6 will examine the roles of CSOs as critics of Japan's ODA policy.

2 A Brief History of International Development CSOs and Government Partnerships[7]

2.1 Civil Society: A Quite New Sector in Japan

Compared to European, North American, and Southern Pacific DAC member countries, Japan does not have a long history of CSOs. As you see in Fig. 1, less than 10% of international development CSOs were established before 1980.

2.2 1980s: Emergence of NGOs and Establishment of Government's Funding Schemes at the End of the Decade

In the early 1980s, the Indochinese refugee crisis led to the establishment of several NGOs such as Japan International Volunteer Center

recommendation, or a less binding guidance A. fter negotiations, DAC members agreed on the adoption as a recommendation.

[7] Chronological review of international development NGOs/CSOs in Japan includes Menju & Aoki (1995), Hirata (2002), Ito (2004, 2007), and Shigeta (2017: Chapter 6 & 7).

(JVC), Shanti International Volunteer Association (SVA: Sotoshu Volunteer Association at the beginning), and Association for Aid and Relief, Japan (AAR). In their early days, most had little interest in policy issues. On the other hand, some social movement groups started to work on international development issues and were very critical of the government's ODA policy. The typical among them was the Pacific Asia Resource Center (PARC). The two kinds of groups had little contact.

In the late-1980s, networks were founded. Kansai NGO Council in the Osaka-Kobe-Kyoto region was established in 1985, and Japan NGO Center for International Cooperation (JANIC), the largest network, in 1987.

Until the mid-1980s, the GoJ paid little attention to the roles of NGOs in international development, and the number of aid agency staff members that had in-depth knowledge about NGOs was little. The early and mid-1980s could be considered as the era of mutual non-engagement between the NGOs and the government.

By the mid-1980s, Italy and Japan were the only two then DAC members which had not established funding schemes for NGOs, although Italian NGOs were able to access funding from then the EC. Because of international criticisms on lack of CSO funding schemes and domestic criticisms on ODA policy especially after the Marcos Scandal in 1986, GoJ started considering partnering with NGOs. In 1989, GoJ finally established the following two funding schemes:

NGO subsidy program for Japanese NGOs (handled by Ministry of Foreign Affairs [MoFA] headquarter).

Grassroots grants program for Southern, International and Japanese NGOs and local governments (handled by local Japanese embassies).

When NGOs learned that GoJ had started considering the establishment of funding schemes for NGOs, the NGO community was split into groups that supported it as an opportunity to scale up their activities and complement the weaknesses of ODA programs, and those that opposed, fearing that NGOs would be co-opted by the government and would lose their independence.

After the funding schemes started, although there had been criticisms on the funding schemes guidelines such as the government's preference of "hardware"—school buildings, clinics, and others that are physically visible—to "software"—capacity building, researches, NGOs became less skeptical that the schemes were means for the GoJ to co-opt NGOs.

2.3 *Booming of NGOs in the 1990s*

The 1990s was the decade of booming of NGOs; fifty percent of the currently active CSOs were established in the decade (Fig. 1). Two factors prompted the expansion of NGOs in terms of the number of groups. First was the establishment of the Volunteer Saving Fund by the then Ministry of Postal and Telecommunications. It was a scheme in which the annual interest of the depositors who joined this program was distributed to NGOs. The founding of this scheme led to the establishment of many NGOs. But after several years, because of the lowering interest rate, the money the ministry could distribute shrunk, and as the result of the privatization of the saving fund, the scheme was abolished in September 2007.

The second factor was the "Kobe effect" (Randel & German, 1997). Many NGOs were active in emergency relief operations and proved their effectiveness at the Great Hanshin Awaji Earthquake in January 1995 and led to the booming of voluntary actions. This "volunteer boom" led to the enactment of the Law on Not-for-Profit Organizations in 1998.[8]

In the 1990s, various schemes for capacity building of Japan's NGOs such as the NGO Consultant Scheme were established. Also important in this decade is the establishment of policy dialogues. In 1996, regular dialogue between CSOs and MoFA (NGO-MoFA Regular Consultation Meeting) was launched. Consultation between JICA and CSOs was started in 1998, and the one with JBIC (then the loans implementation agency) in 2001.

2.4 *After 2000*

The number of newly established CSOs decreased in the 2000s and 2010s (Fig. 1). Many of them were established by the younger generation. Another trend of the international development CSOs that emerged in the 2000s and 2010s was strengthened partnership with the business sector and trade unions.

In 2000, the GoJ, the business community, and several NGOs established the Japan Platform which "coordinates the efforts of NGOs, the business community, and the government, all of which participate as equal

[8] Until this law was enacted, most of Japanese CSOs were not incorporated.

partners"[9] and emphasized the "All Japan approach" in emergency relief. NGOs were again split into those who supported and joined this as an innovative multi-sector initiative to work on emergency relief and those who were skeptical that NGOs that joined would be forced to align with the government's positions, especially in conflicts, and could undermine NGOs' independence. Skeptical groups did not join the Platform.

In 2002, a new scheme "Grant Assistance for Japanese NGO Projects" was established for NGOs with proven records. While it is project-based funding, it is multi-year funding, and NGOs could use up to 5% (currently 15%) for administration.

In the twenty-first century, there is consensus that the Millennium Development Goals (MDGs) and the Sustainable Development Goals (SDGs) are the common goals for both the government aid agencies and CSOs. But as I will describe in the section on advocacy, the views on means for achieving the goals have continued to be different; while the government emphasized the importance of growth and allocated big proportion of bilateral ODA to economic infrastructure, CSOs have rejected the growth-centered view and called for more focus on direct support for poverty reduction and social development. More important is that CSOs have been critical of the recent trends in the government's ODA policy; instrumentalization of ODA for political and commercial purposes, as I will describe later in the section on CSOs' advocacy.

There have not been major changes in both funding schemes and dialogues. After the amalgamation of JICA and JBIC in 2008, dialogue between the "new JICA" and CSOs continued.

3 Japanese International Development CSOs at a Glance

Here, I will briefly describe the current trends of international development CSOs in Japan, based on the *NGO Data Book* (JANIC, 2016) by JANIC.

[9] Japan Platform website (https://www.japanplatform.org/E/about/jpf.html: Accessed 27 February 2021).

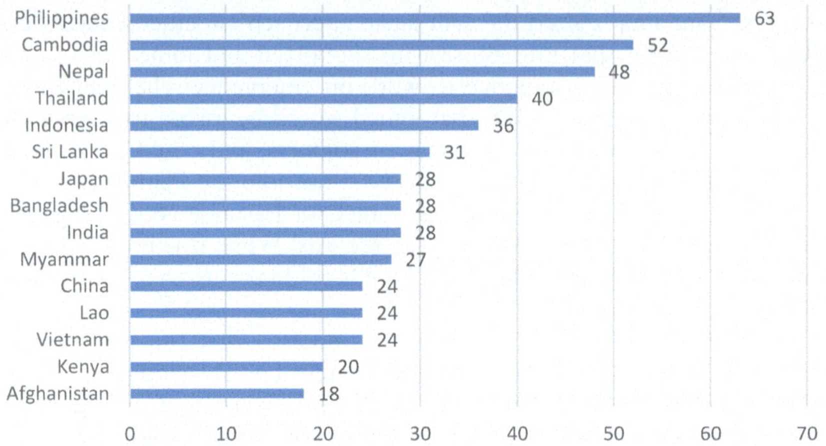

Fig. 2 Top 15 countries Japanese CSOs are active (n = 430) (*Source* JANIC, 2016)

3.1 *Geographical and Sectoral Trends*

If we look at the top fifteen countries in which Japanese CSOs are active (Fig. 2), all countries except Kenya are in Asia. Readers might wonder why Japan is ranked seventh. It is estimated that at the time the data was collected, many Japanese CSOs worked on the reconstruction of communities damaged or affected by the Great East Japan Earthquake and the tsunami on March 11, 2011.

Sectorally, the global trend of CSOs' emphasis on social development sectors such as education, health, and rural development also applies to Japan. But there is a stronger emphasis on education. Compared to CSOs in Europe and North America, fewer CSOs work on human rights and gender (Fig. 3).

3.2 *Types of Activities*

Although more than half of Japanese CSOs provide financial support in partner countries, the percentage of CSOs that dispatch personnel to partner countries is quite big; nearly half (Fig. 4). This is a reflection of the fact that despite their small size, Japanese CSOs have a preference for directly administrating or implementing projects and programs. Part of

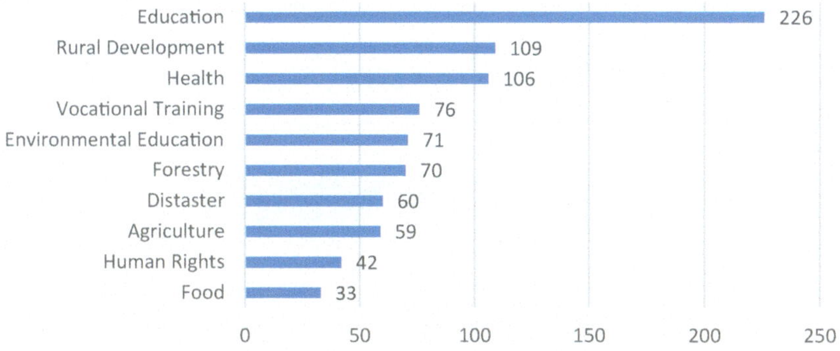

Fig. 3 Sectors Japanese CSOs work on (n = 430) (*Source* JANIC, 2016)

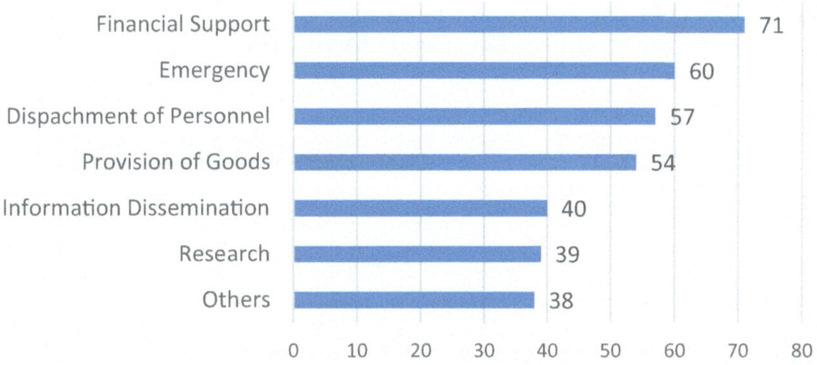

Fig. 4 Types of activities (Overseas) by Japanese CSOs (n = 124) (*Source* JANIC, 2016)

the reason, as we shall see later, is that the GoJ requires Japanese CSOs' involvement in the implementation of headquarter-funded projects.

Domestically, while nearly 60% provide information to the Japanese society, only 40% work on global education, and only 20% work on research (Fig. 5), meaning that CSOs working on transformation of attitude of Japanese public and policy advocacy is quite limited.

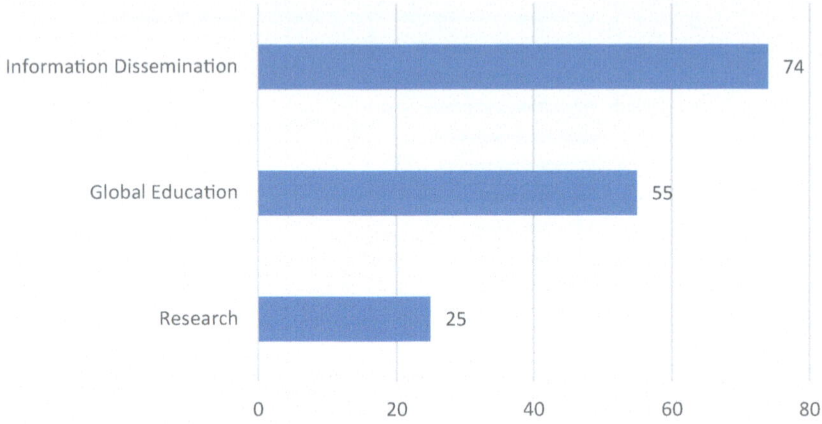

Fig. 5 Types of Domestic Activities by Japanese CSOs Related to Development Education and Advocacy (n = 124) (*Source* JANIC, 2016)

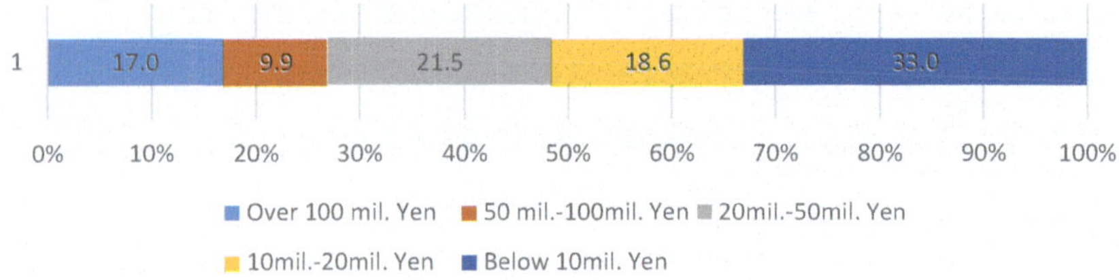

Fig. 6 Annual budget of Japanese CSOs (%) n = 312 (*Source* JANIC, 2016)

3.3 Many Groups are Small in Terms of Annual Budget

While 17% of international development CSOs in Japan have an annual budget of over 100 million Yen, over half of the groups work annually with less than 20 million Yen (Fig. 6). Among the largest ones with over 100 million Yen budget are the Japanese affiliates of International CSOs (ICSOs) such as Medicines sans frontiers, Plan, World Vision, and Save the Children.[10] In reality, I guess that there are 200 to 300 more very

[10] Not all ICSOs have been successful in Japan. Oxfam Japan, founded in 2003, was closed in 2018 because of financial difficulty.

Table 1 Aid by CSOs of DAC member countries

	Aid by CSOs (US$ mil.) (b)	*Population (mil.)*
Canada	2,893	37.1
Denmark	353	5.8
Germany	2,239	83.1
Greece	320	10.5
Ireland	533	4.8
Italy	61	60.6
Japan	622	127.2
Korea*	420	51.2
Netherlands*	114	17.1
New Zealand*	150	4.7
Portugal	10	10.3
Switzerland	676	8.5
United States	36,291	327.1

* 2017

Source OECD (2020a), Population: UNFPA

small CSOs (with an annual budget of around 1 million Yen and run entirely by volunteers), which are not surveyed by JANIC.

OECD estimates that Japan's "aid by CSOs" in 2018 was US$625 million, which is much less compared to the United States, Canada, and Germany, or around the same as Ireland or Switzerland with much less population (Table 1).[11]

4 CSOs as Partners of ODA Policy

4.1 Current Funding Schemes

During the 30 years' history of partnering with CSOs, the government support for CSOs scaled up in terms of both the amount of money and the number of schemes. Currently, there are the following funding schemes by MoFA and JICA;

- Grant Assistance for Japanese NGO Projects: In FY 2019, a total of about 4.3 billion Yen was provided for 86 projects by 45 CSOs.
- NGO Project Subsidies: for studies and evaluations;

[11] In OECD (2020a), data on "aid by CSOs" was available for only a limited number of countries.

- NGO Consultant Scheme: funding for regional consultants disseminating on development issues;
- NGO Intern Program: supporting young people;
- NGO Overseas Study Program: assisting NGO staffs' overseas training;
- NGO Study Program: support for studies on specific topics;
- Grant Assistance for Grass-Roots Human Security Projects: handled by local Japanese embassies and consulates;
- JICA Partnership Program.

4.2 Lack of CSO-Specific Policy Framework

As strengthening partnerships with CSOs is explicitly mentioned in the Development Cooperation Charter (MoFA, 2015), Japan is among the 22 DAC members with a policy on civil society (OECD, 2020b: 60–63). Japan is not among the sixteen members that have a policy framework specific to civil society partnership while Korea is, as the result of the launching of a policy framework on CSO partnership in 2019.

According to the Development Cooperation Charter:

> Partnerships with the civil society in and outside of Japan, including NGOs, civil society organizations (CSOs), and private foundations, are important both for greater cooperation effectiveness and for the equitable and stable development of the recipient countries as they can accurately assess varying views and needs on the ground and take timely flexible actions. With this recognition, the government will strategically strengthen partnerships with NGOs/CSOs, including reinforcing their participation and collaboration in development cooperation. From this standpoint, the government will support excellent development cooperation projects of Japanese NGOs/CSOs and their capacity development. In this regard, the Ministry of Foreign Affairs and JICA will focus on developing human resources and systems in the social development sector. (MoFA, 2015)

It looks like GoJ's objectives for working with CSOs include implementing programs related to service delivery and enhancing Japanese CSOs' capacity. The paragraph does not include strengthening pluralistic and independent civil society in partner countries or promoting democratization and human rights as objectives for partnering with civil society. There is also no mention of CSOs as independent development actors in their own right. It also does not address the issues around civic space and CSO enabling environment.

DAC members, such as Canada, Sweden, Norway, Finland, Denmark, Ireland, and EU which have had a long tradition of partnering" with CSOs, have policy frameworks or strategies regarding their partnerships with CSOs (Global Affairs Canada, 2017; EU, 2017; Irish Aid, 2008; Ministry of Foreign Affairs of Denmark, 2014; Ministry of Foreign Affairs of Finland 2017; Norad, 2018; Sida, 2019; Swedish Ministry of Foreign Affairs, 2016). They explicitly state that CSOs are key actors in promoting democracy, democratic and participatory development, and human rights and human rights-based approaches to development (HRBA).

4.3 Tiny Percentage of ODA for CSOs

As I said at the beginning of this chapter, while 15.1% of the total ODA of the DAC members are for CSOs, the figures are extremely low at 1.7% for Japan (and 2.1% for Korea). In 2018, Greece had no aid for CSOs, and Japan and Korea are the second and the third from the last among the DAC members in terms of the proportion of ODA for CSOs (Fig. 7).

Several reasons for small allocation for CSOs could be named. First, considering that many Japanese CSOs are small as we saw in Table 1, there is a problem with the absorptive capacity on the CSO side, which I mentioned earlier. Related to this is the issue of avoiding over-dependence on government funding; if GoJ rapidly increases its financial support to CSOs, the dependence of CSOs on government funding would become too big. Large dependence of CSOs on government funding could lead to CSOs being "tamed" by the government, becoming executing agencies of government policies and priorities rather than independent actors in their own right, and vulnerable to changes of the government or government policies as happened in Canada and Finland (Takayanagi, 2020).

Second, this is a reflection of the fact that Japan emphasizes the public sector as a channel of delivery of its aid, channeling 85.8% through the public sector. In comparison, the figure for DAC members in total was 50.2% in 2018.[12]

Third, in terms of sectoral allocation, Japan has emphasized aid for economic infrastructure, allocating of 57.3% its ODA in 2018. Most

[12] OECD, Development Co-operation Profiles. Korea is similar to Japan in this respect, channeling 79.6% of its aid through the public sector.

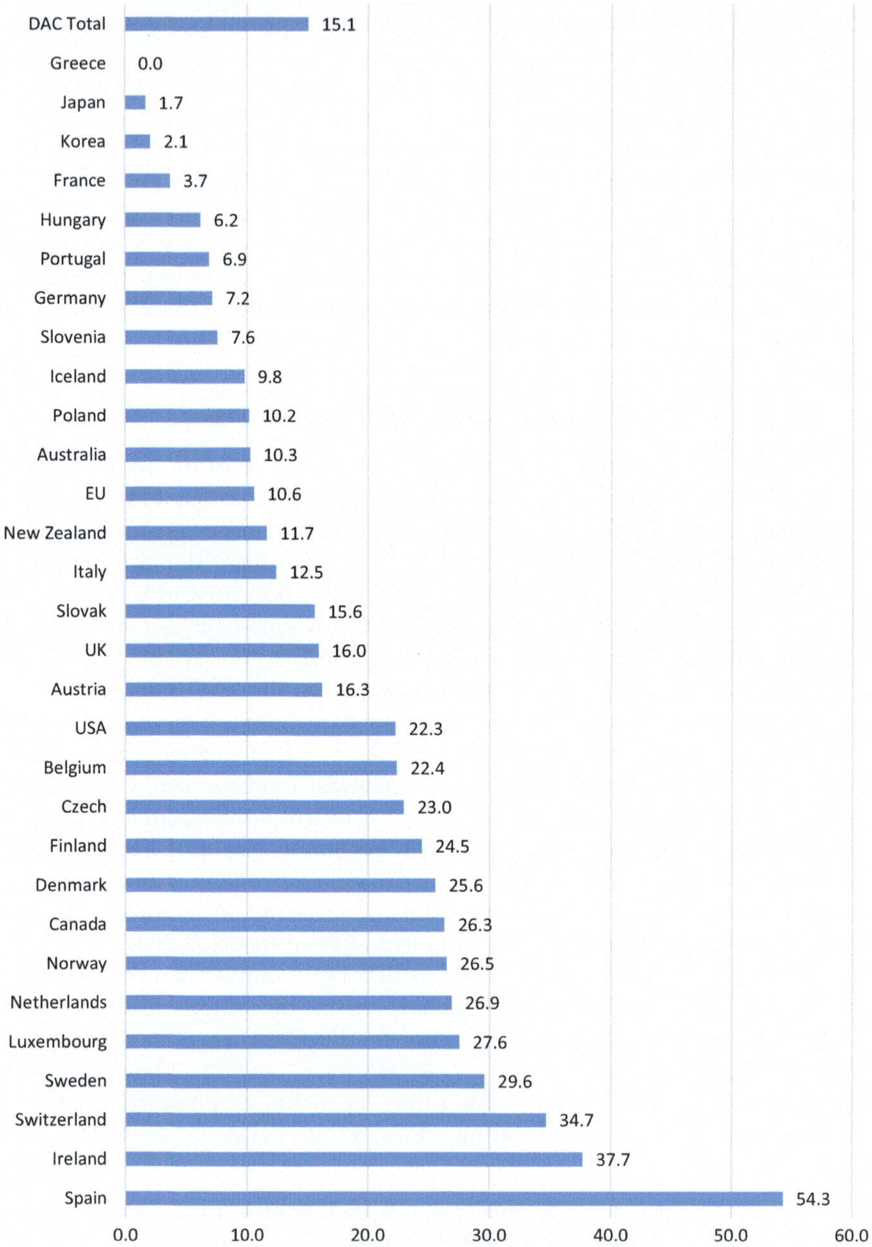

Fig. 7 Aid for CSOs in 2018 (%) (*Source* OECD [2020a])

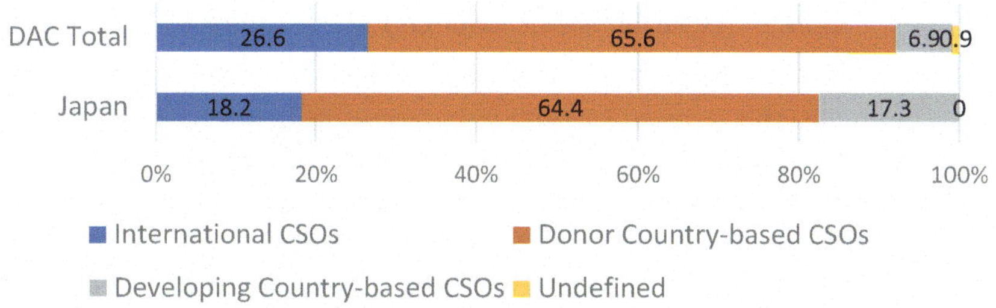

Fig. 8 Allocation by types of CSOs (2018, %) (*Source* OECD [2020a])

DAC members, especially those with good records in aid for CSOs, have emphasized social infrastructure and humanitarian assistance in their aid policies and allocation. These sectors are where CSOs' works are concentrated. Japan's emphasis on sectors in which CSOs don't (and can't) work intensively is another reason why Japan's proportion of aid for CSOs is so low.

4.4 Support for Partner County-based CSOs

If we look at the allocation to different types of CSOs, Japan's distribution to partner country-based CSOs is relatively larger, while to ICSOs is smaller (Fig. 8). This is probably because of the Grant Assistance for Grass-Roots Human Security Projects.

However, to get headquarter funding, Japanese CSOs must be involved in project implementation—projects and programs implemented by partner country CSOs are ineligible for government funding.[13] This is probably because of the GoJ's policy of "Visible Japanese Development Cooperation."[14] This policy is incompatible with the objective of promoting democratization and human rights in partner countries through strengthening pluralistic civil society.

[13] This is explicitly stated in the guideline of the Grant Assistance for Japanese NGO Projects.

[14] According to MoFA, "Visible Japanese Development Cooperation" is "a means of ensuring that people in developing countries recognize that the ODA is implemented by Japan" (MoFA, 2015: 19).

4.5 Challenges in CSO-GoJ Partnership

Summing up this section, I would name some challenges for GoJ based on the recent OECD documents on CSO partnerships and the DAC Peer Review in 2019/2020.

5 Partnership Policy Framework

The Recommendation and an action point in the OECD study say that all members should have a policy document on civil society or CSOs that;

> a) articulate objectives for working with a diverse range of civil society actors both as independent development actors in their own right and as implementing partners;
> b) aim to strengthen local ownership and an inclusive and independent civil society in partner countries or territories;
> c) take into account contextual risks or opportunities for civil society and civic space; and
> d) integrate these policy or strategy positions into wider development co-operation and humanitarian assistance policies or strategies. (OECD, 2021)

GoJ should consider a CSO partnership policy framework that explicitly recognizes CSOs as independent development actors in their own right and that strengthening civil society is essential for promoting democratization and human rights. Also, such a document should address the issue of civic space and CSO enabling environment in partner countries.

6 Funding Mechanisms

The DAC Peer Review Report in 2020 recommended that Japan should provide "greater institutional support to civil society organisations in Japan and partner countries as strategic partners and development actors in their own right (OECD, 2020c)."

I mentioned that the proportion of ODA for CSOs is tiny compared to other DAC members in Japan. Enhancing support for CSOs in Japan is indispensable. But considering the size (annual budget, number of staff members, etc.) of Japanese CSOs, it is necessary to consider the absorptive capacity of CSOs and to avoid CSOs' over-dependence on government funding.

Possible other solutions include increased funding to international and partner country-based CSOs. The Recommendation and an action point in the OECD study call for increased financial support for partner country-based CSOs (OECD, 2021, 2020b: 111). Increasing the Grant Assistance for Grass-Roots Human Security Projects will require strengthened capacity at local embassies and consulates. Another possibility Japan could think about is joining or establishing multi-donor pooled funds for supporting CSOs with other DAC members, which is an idea proposed in the OECD study (Ibid.: 109).[15]

In addition, an important challenge for GoJ is how to make the existing funding schemes and mechanisms more flexible. GoJ's funding for CSOs is basically project-based rather than program-based. Core funding[16] has not been accepted. GoJ prefers "hard wares" such as school buildings and clinics, while "soft wares" such as capacity building or research are ineligible or not preferable. Possible measures include introducing program and core funding and amending the funding rules and practices to enable financing for "software" like capacity building or policy research. Another important thing, as mentioned in the Recommendation and the OECD study (OECD, 2021, 2020b: 113–114), is to reduce administrative requirements for CSOs. Also, GoJ should reconsider its policy of requiring Japanese CSOs to be involved in the implementation of headquarter-funded projects.

[15] However, it must be understood that pooled fund is not a "magic bullet" for supporting civil society. There is a possibility that increase of donors' pooled funds could reduce funding choices for CSOs, narrowing CSOs' interaction with DAC members and, depending on how pooled funds are managed, could become donor-driven (2020b; OECD, 2011).

[16] Core funding is "aid funds to CSOs to support the development activities of the CSOs themselves" (OECD 2011: 11). It is often multi-year support for the program of the recipient CSOs. It is basically welcomed by CSOs as the responsibility for managing the fund would rest with CSOs, giving them more flexibility. Because core funding is often multi-year support, it gives CSOs predictability (OECD, 2011; 2020b). However, it also has disadvantages such as possibility of recipient CSOs becoming over-dependent on official funding, and difficulties in demonstrating results and effectiveness (OECD, 2020b: 36).

7 Policy Dialogue

7.1 Policy Dialogue Between CSOs and GoJ

According to the OECD study, all DAC members consult with CSOs on their policies, and twenty members hold regular (prescheduled) dialogues or consultations (OECD, 2020b: 84–87). Why are dialogue and consultation critical? According to the OECD study;

> Dialogue and consultation with CSOs are integral to members' commitments to openness and transparency. They also foster better and potentially more coherent development co-operation, foreign policies and programmes by taking advantage of CSOs' knowledge, expertise and experience including on civic space challenges. (Ibid.: 40-41)

In Japan, regular dialogue between CSOs and MoFA was launched in 1996 as already mentioned. Currently, a general meeting is held once a year, and the two subcommittees, the "ODA Policy Council," which deals with aid policy and the "Partnership Promotion Committee" which focuses on MoFA funding schemes and other support mechanisms, are both held three times a year. At the headquarter level, NGO-JICA Consultation Meeting has been held four times a year.

Since 2002, "NGO-Embassies ODA Consultation Meetings" have been held in partner countries where many Japanese CSOs operate (MoFA, 2019). However, this is between field staff members of Japanese CSOs and MoFA/JICA. Japan has not institutionalized dialogues with partner countries' CSOs.[17]

8 Challenges for CSOs-GoJ Dialogue

The Recommendation call for a "more structured, institutionalised, inclusive and accessible" (OECD, 2021) policy dialogue, especially at the partner country level. In this context, whether policy dialogue has been meaningful in Japan is a critical question. This is because CSOs "cannot refrain from wondering how effectively their voices have been heard in terms of substantially changing ODA policies and ideologies (Ohashi, 2016: 339)." Now that it is almost 25 years since policy dialogue and

[17] According to the OECD study, only seven DAC members hold regular policy dialogue with partner country CSOs (OECD, 2020b).

consultation have been institutionalized in Japan, the time has come to examine what were the changes that dialogue and consultation brought about.

What is missing in the case of Japan is the institutionalized dialogue with partner country's CSOs and joint dialogue with other DAC members' CSOs.

9 CSOs as Critics of ODA Policy

9.1 *Japanese CSOs' Advocacy on Aid Policy: An Overview*

The history of advocacy on GoJ's aid policy by Japanese CSO was recently comprehensively reviewed by Ohashi (2016). His review covers the period from the early 1980s, when CSOs' advocacy on aid policy started, to the mid-2010s, when CSOs worked on advocacy regarding amendment of the ODA Charter, including the change of the document's name to the Development Cooperation Charter in 2015. Ohashi's conclusion includes the followings (Ibid.: 341):

- There were some cases that CSOs' advocacy was successful in terminating or freezing specific ODA projects or improving the terms and conditions of CSO funding by MoFA. But these were "rather minor within Japan's ODA as a whole."
- Some global campaigns such as Jubilee 2000 were successful in Japan, while the campaign by GCAP (Global Call to Action Against Poverty) in 2005 "demonstrated how advocacy is not always well understood in Japanese society."[18]
- CSOs in Japan have always advocated for a framework legislature that states the objectives and principles of GoJ's aid policy. But it has not been successful "as the administration has no intention of giving up its monopoly over ODA governance."

To keep duplication with Ohashi's review to the minimum, I will focus on three topics related to ODA policy of the GoJ19; how CSOs in Japan views the "Japan's Aid Model," follow-up of the CSOs' advocacy on the Development Cooperation CharterDevelopment Cooperation

[18] In GCAP's campaign, benefit from selling the white bands was announced to be supporting CSOs' research and advocacy works, but many people who bought the white bands thought that it would go to projects that work on poverty alleviation.

Charter, Japan and what were the voices of Japanese CSOs at the DAC Peer Review in 2019/2020.

9.2 CSOs' Criticisms on "Japan's Aid Model" through Contribution to The Reality of Aid

Globally speaking, as critics of aid policy CSOs have tended to articulate the followings;

- In terms of aid volume: All DAC members should increase their aid to meet the internationally agreed goal of 0.7% of GNI.
- In terms of aid objectives: global and enlightened interests over self-interests (political/diplomatic/strategic and economic/commercial). Aid should not be instrumentalized for political or commercial objectives.
- In terms of development visions: As social movement raising critical voices on development (aid) policies of the governmental agencies and the works of the private sector have questioned growth-oriented visions and especially recently have been advocates of human rights-based approaches (HRBA)

The Reality of Aid (RoA) is a network of CSOs from both donor and partner countries that has been critically analyzing aid policies and practices. It was started in 1993 by Northern NGOs as an "alternative DAC report." Since the mid-1990s there have been contributions from Southern NGOs. At the beginning of the twenty-first century, the secretariat was moved to the South and has been in the IBON Foundation in Manila. RoA has published its global report every year, or since 2002, every other year.

JANIC has been contributing to the Japanese country reports since the start of the RoA.[19] JANIC writers have repeatedly written that;

- Aid should be increased, and a timeframe for achieving the 0.7% ODA/GNI target should be set;
- More aid should be allocated to sub-Saharan Africa;
- There has been too little allocation for social infrastructure (health, education, government and civil society, etc.) and too much allocation for economic infrastructure;

[19] The newest report at the time of writing this chapter is Takayanagi (2021).

- The proportion of grants should be increased;
- Increased transparency.

It has been said that geographical emphasis on Asia, sectoral emphasis on economic infrastructure, and heavy use of loans have been the characteristics of Japan's ODA, sometimes described as "Japan's Aid Model." RoA's chapters on Japan have always been critical about the "Japan's Aid Model."

9.3 CSOs and the Amendment of the ODA Charter to the Development Cooperation Charter

One significant recent development in Japan's aid policy was the amendment of the ODA Charter, including changing its name to the Development Cooperation Charter (other chapters this volume?). The process started in March 2014 with the establishment of an experts' panel, and the new Charter was announced in 2015. Although partly covered in the work by Ohashi (2016), I will summarize Japan's CSOs' advocacy works regarding this significant change in the basic framework of GoJ's aid policy.

JANIC, with the collaboration of other CSOs, issued statements during the process and after the announcement of the new policy (JANIC, 2014, 2015). CSOs did not reject the new Charter altogether. They acknowledged several positive aspects of the new Charter including;

- "promoting women's participation" as the main principle;
- strengthening partnerships with CSOs as one of the "implementation arrangements";
- noting the internationally agreed 0.7% ODA/GNI target (however, there was no timeframe)
- emphasizing the importance of development education and public engagement.

But CSOs expressed their concerns, especially on the following three points. The first two are concerns regarding strengthened instrumentalization of ODA for "national interests."

- Linking ODA and Security Policies Abe government at the beginning said that the Charter revision is closely related to their security strategy and wanted to step into security-related projects (especially

in the Philippines and Vietnam according to media reports). After debates, the new Charter allows supporting welfare and humanitarian works by military personnel, considered case-by-case.
- (Re)commercialization of Aid: At the beginning of the process was related to the revitalization strategy, "Dialogue and collaboration based on Japan's experience," "cooperation that takes advantage of Japan's strength," "proactively presenting proposals": possibly connect commercial interests and undermine developing country ownership.
- Growth-oriented views: 1993 and 2002 Charters put both poverty reduction and growth as priorities. The new Charter describes "Quality growth and poverty reduction through such growth." It puts more emphasis on growth (although they say, "quality growth," which, as a result, shares similarities with SDG 8) and considers poverty eradication as a result of growth.

9.4 DAC Peer Review and CSOs in Japan

The most recent effort by CSOs in Japan was its contribution to the DAC Peer Review. During the headquarter visit of the review team (EU, Italy, and the OECD Secretariat) in November 2019, there was a meeting with CSOs. Coordinated by JANIC, Japanese CSOs prepared their shadow report with contributions by seven CSOs (JANIC, 2019). JANIC summarizes the key messages of the report as the following.

- The CSO community is critical to the new Development Cooperation Charter's emphasis on quality growth. Instead, "fair distribution" and "poverty eradication" should be promoted. Also, instead of the Charter, there should be a legal framework be made by the legislature to define the visions, missions and objectives of Japan's ODA.
- Japan's heavy use of loans and emphasis on economic infrastructure should be fundamentally reconsidered.
- More attention should be paid to negative impacts such as human rights violations and environmental issues caused by some ODA projects, particularly in Asian countries.

9.5 Challenges for CSOs in Japan's Advocacy

I have described that CSOs in Japan, as critics of Japan's ODA policy have been critical to "Japan's Aid Model." I would name some challenges for Japan's CSO community in terms of advocacy. Considering that while CSOs are critical with GoJ's aid policy but GoJ and CSOs share interest in enhancing public engagement in international development, how CSOs could gain increased public support is important. Related to this is that CSOs in Japan have to overcome is that not all of their constituencies share the understanding that policy research and advocacy are important roles for CSOs. How to strengthen CSOs' capacity in policy research is another challenge. Especially that Japan's CSOs are small, both financial and human resources for policy research have been quite limited. Also, the number of CSOs that work on research is small as shown in Fig. 5.

10 Conclusion

In this chapter, I described how CSOs in Japan and GoJ have partnered in international development, how policy dialogues have taken place and how Japanese CSOs have played the roles as critics of GoJ's aid policy. My conclusion is that while GoJ's partnership has enhanced in the past 30 years, financial support for CSOs is still limited, and it is not explicit whether CSOs are recognized as independent development actors in their own right, and that promoting democratization and human rights through strengthening pluralistic and independent civil society in partner countries is essential.

I identified several challenges, referring to the Recommendation and the action points in the recent OECD study on DAC members' partnership with CSOs.

- GoJ should have a CSO policy framework beyond just a few paragraphs in the Development Cooperation Charter. The new framework should explicitly recognize CSOs as independent development actors in their own right and CSO roles not only in service delivery but also in promoting democratization and human rights.
- GoJ should increase financial support for CSOs and introduce new ways of funding such as core funding and joint pooled funding with other DAC members.
- Although policy dialogue between CSOs and GoJ has been institutionalized at the headquarter level, it should be examined what changes have been brought about through dialogue, to make it even more meaningful.
- Strengthening partnership with partner countries-based CSOs is a critical key challenge for GoJ. While the proportion of funding for partner country CSOs among Japan's aid for CSOs is bigger compared to the DAC total, GoJ's objective for partnering with CSOs does not include strengthening civil society in partner countries

and has not institutionalized policy dialogue with partner country CSOs.

I should add a few challenges for CSOs in Japan which are cross-cutting for partnership with GoJ, policy dialogue, and advocacy. First is strengthening the constituency for CSOs. As shown in Table 1, the amount of CSOs' aid is very small. CSOs in Japan are recently struggling with fundraising from the public. A significant reason for this is the decreasing and aged population. How to strengthen their constituency is becoming more of a challenge for CSOs in Japan. Increased government funding without diversified funding sources would make CSOs mere implementers of government's policies and priorities instead of independent development actors in their own right.

Second, CSOs must improve their effectiveness, part of which is transparency and accountability.[20] The OECD study calls for improved transparency and accountability of both CSOs and DAC members in Action Point 3.1.9 (OECD, 2020b: 115–116).

References

Banks, N., & Hulme, D. (2012). *The role of NGOs and civil society in development and poverty reduction* (Brooks World Poverty Institute Working Paper).

Banks, N., Hulme D., & Edwards, M. (2013). Epilogue: Still too close for comfort? In Hulme, D. & Edwards M. (Eds). *NGOs states and donors: Too close for comfort?*, 2nd ed. Palgrave Macmillan.

Bebbington, A., Hickey, S., &Mitlin, D. (eds). (2008). *Can NGOs make a difference: The challenge of development alternatives*. Zed.

Edwards, M. (2011). *Introduction: Civil society and the geometry of human relations, Edwards* (M). Oxford University Press.

Edwards, M. (2020). *Civil society* (4th ed.). Polity.

EU. (2017). *Report on EU engagement with civil society*.

Global Affairs Canada. (2017). Canada's policy for civil society partnerships for international assistance: A feminist approach. https://www.international.gc.ca/worldmonde/issues_development-enjeux_developpement/priorities-priorites/civil_policypolitiquecivile.aspx?lang=engAccessed August 25, 2020

[20] The reason I say transparency and accountability are part of the effectiveness agenda is that among the Istanbul Principles for CSO Development Effectiveness, which are considered to be one of the guiding principles for partnering with CSOs by some DAC members such as Canada and Sweden, is improving transparency and accountability (Open Forum for CSO Development Effectiveness, 2011).

Hirata, K. (2002). *Civil society in Japan: The growing role of ngos in Tokyo's aid and development policy*. Palgrave Macmillan.

Howell, J., & Pearce, J. (2001). *Civil society and development: A critical exploration*. Lynne Rienner.

Hulme, D., & Edwards, M. (Eds.). (1997). *NGOs states and donors: Too close for Comfort?* Macmillan.

Irish Aid. (2008). *Civil society policy*. https://www.irishaid.ie/media/irishaid/allwebsitemedia/20%20newsandpublications/publicationpdfsenglish/civil-society-policy.pdf. Accessed August 25, 2020.

Ito, M. (2004). Nihon no Kokusai Kyoryoku NGO no Rekishi to Network-ka no Nagare (History of International Cooperation NGOs and Networking in Japan). In Imata, K. & Harada,K. (Eds.), *Renzoku Kogi Kokusai Kyoryoku NGO (Lecture Series on International Cooperation NGOs)*. Nihon Hyoron-sha.

Ito, M. (2007). Nihon Shakai to NGO: Kokkyo wo Koeru Shimin Soshiki (Japanese Society and NGOs: Transnational Civil Society Organizations. In Umahashi, N. & Takayanagi, A. (Eds.), *Global Mondai to NGO, Shimin Shakai (Global Issues and NGOs/CSOs)*. Akashi Shoten.

JANIC. (2014). *Kokusai Kyoryoku NGO ni Yoru ODA Taikou Minaoshi 10 no Teigen* (Ten Proposals by International Cooperation NGOs on the Amendment of the ODA Charter).

JANIC. (2015). *Kaihatsu Kyoryoku Taikou Kakugi Kettei ni Taisuru Kokusai Kyoryoku NGO no Kinkyu Seimei* (Urgent Statement by International Development Cooperation NGOs on the Cabinet's Approval of the Development Cooperation Charter).

JANIC. (2016). *NGO data book*.

JANIC. (2019). *Civil Society Report for OECD-DAC Peer Review: Japan*.

Kaldor, M. (2003). *Global civil society: An answer to War*. Polity.

Kim, S., & Hong, J. (2022). The government-civil society relationship in Korean international development cooperation from a historical perspective. This Book.

KOICA .(2019). *Policy framework for government-civil society partnership in international development cooperation*.

Lewis, D., Kanji, N., & Themudo, N. (2021). *Non-governmental organizations and development* (2nd ed.). Routledge.

Menju, T., & Aoki, T. (1995). NGOs: Japan. In Yamamoto, T. (Ed.), *Emerging civil society in the Asia Pacific community*. Japan Center for International Exchange.

Ministry of Foreign Affairs of Denmark. (2014). *Policy for Danish support to civil society*. https://amg.um.dk/policies-and-strategies/policy-for-support-to-danish-civil-society/. Accessed 25 August 2020.

Ministry for Foreign Affairs of Finland. (2017). *Guidelines for civil society in development policy*

MoFA. (2015). *Development cooperation charter: For peace, prosperity and a better future for everyone.*
MoFA .(2019). *White paper on development cooperation 2018* (English Version).
Norad. (2018). *Norad's support to civil society: Guiding principles.* https://norad.no/contentassets/396cdc788c09405490a96adce80ac040/norads-support-tocivil-society-guiding-principles.pdf. Accessed 25 August 2020.
OECD. (2011). *How DAC members work with civil society organisations: An overview.*
OECD. (2012). *Partnering with civil society: 12 lessons from DAC peer reviews.*
OECD. (2020a). *Aid for civil society organisations: Statistics based on DAC Members' reporting to the Creditor Reporting System database (CRS), 2017–2018.*
OECD. (2020b). *Development assistance committee members and civil society.*
OECD. (2020c). O*ECD development co-operation peer reviews: Japan 2020.*
OECD. (2020d). *DAC high meeting communique 2020.*
OECD. (2021). *DAC Recommendation on Enabling Civil Society in Development Co-operation and Humanitarian Assistance.*
Ohashi, M. (2016). NGOs and Japan's ODA: Critical views and advocacy. In Kato, H., Page, J. & Shimomura, Y. (Eds.), *Japan's development assistance: Foreign aid and the post-2015 agenda.* Palgrave Macmillan.
Open Forum for CSO Development Effectiveness. (2011). *Siem reap consensus on the international framework for CSO development effectiveness.*
Randel, J., & German, T. (1997). Japan. In Smillie, I. & Helmich, H. (Eds.), *Stakeholders: Government-NGO partnerships for international development.* Earthscan.
Shigeta, Y. (2017). *Gegido suru Global Shimin Shakai (Global Civil Society in Turbulence).* Akashi Shoten.
Sida. (2019). *Development cooperation with focus on adaptivity and trust.* https://www.sida.se/%20English/press/current-topics-archive/2019/development-cooperation-withfocus-on-adaptivity-and-trust/ . Accessed 25 Aug 2020.
Swedish Department of Foreign Affairs. (2016). *Strategy for support via Swedish civil society organisations for the period 2016–2022.*
Takayanagi, A. (2020). CSO-government partnership: Lessons from the canadian experience. *The Journal of Global and Intercultural Studies*, 22.
Takayanagi, A. (2021). Increases in Japan's ODA/GNI ratio: Should we celebrate It? In B. Tomlinson (Ed.), *The reality of aid 2020/21: Aid in the context of conflict, fragility, and the climate emergency.* IBON International.
Tvedt, T. (2006). The international aid system and the non-governmental organisations: A new research agenda. *Journal of International Development*, 18.
Van Rooy, A. (Ed.). (1998). *Civil society and the aid industry.* Earthscan.

CHAPTER 10

The Government–Civil Society Relationship in Korean International Development Cooperation from a Historical Perspective

Sung Gyu Kim and Jiyoung Hong

1 Introduction

The discourse on the role of development civil society organizations (CSOs) and their importance has been quite active. In donor countries,

This work was supported by the Ministry of Education of the Republic of Korea and the National Research Foundation of Korea (NRF-2018S1A3A2075117). It was published in International Development and Cooperation Review 13(3).

S. G. Kim
Center for Sustainable Development, Korea University, Seoul, South Korea
e-mail: sungkim0915@gmail.com

J. Hong (✉)
Research Institute of North Korea and Northeast Asia Development, Korea Eximbank, Seoul, South Korea
e-mail: jyhong@snu.ac.kr

H.-j. Kwon et al. (eds.), *International Development Cooperation of Japan and South Korea*,
https://doi.org/10.1007/978-981-16-4601-0_10

government–civil society partnerships have been established since the 1960s. For example, bilateral aid agencies in the US, Germany, Sweden, Canada, and Norway are active and have begun to cooperate with CSOs. In the 1990s, government partnerships with CSOs were institutionalized (Lewis & Kanji, 2009). In the 2000s in particular, the international development community officially recognized CSOs as independent actors and important aid agents for development at the Third OECD High-Level Forum (the HLF-3 in 2008). The role of CSOs has been further emphasized as private finance has become more important in development cooperation, and value-oriented agendas such as environment, democracy, gender, and human rights have been included as development issues since the launch of the Sustainable Development Goals (SDGs).

Korea received foreign aid as a poor country after the Korean War. However, through the process of rapid economic growth and democratization during the 1970s–1980s, the country has grown into a stable democracy and the world's 10th largest economic power. Korea's international development activities have grown dramatically over the past 30 years, and as of 2019, the size of Korea's official development assistance (ODA) was estimated at USD 2.52 billion, ranking 15th among the OECD Development Assistance Committee (DAC) member countries. In addition to the expansion of the ODA scale, the increase in private development assistance played an equally significant role.

CSOs and their partners have been conducting relief work for Korea since the Korean War. In the 1990s, as Korea's economic performance and living conditions improved, these organizations in Korea began carrying out international development projects abroad. During this same time period, development organizations working on international development were established in Korea. In the 2000s, development CSOs increasingly focused on service delivery. Around 2010, when Korea joined the DAC, qualitative changes were observable in terms of the composition of the development. CSOs were diversified, and the scope of their activities expanded. In short, foreign aid activities and Korea's development CSOs increased rapidly within a short period of time compared to other existing donor countries. Particularly in the early stages of such activities in the 1990s and 2000s, the Korean CSO role as an implementor assisting the government in carrying out ODA was highlighted. However, development CSOs are currently proposing various agendas from the perspective of civil society, and some organizations are leading the discourse at the domestic and international levels.

In 2019, the Korean government established the *Policy Framework for Government–Civil Society Partnership in International Development Cooperation*, which serves as an institutional basis for development CSOs to participate in the policy-making process. In order to understand how the government and CSO cooperation have been strengthened and finally become institutionalized in the field of international development cooperation in Korea, it is important to analyze the role of development CSOs and the government–civil society relationship by period. Despite the importance of this role, little research has been conducted on the role of Korean development CSOs (Kim, 2017; Kim & Hong, 2014; Lee, 2012; Sohn, 2015). In particular, there are no existing studies on the process of changes in the government–civil society relationship.

In this context, this study aims to examine how the relationship between the government and civil society has changed from a historical perspective. Additionally, as this relationship is closely linked to the role of CSOs, it is necessary to closely examine how development civil society came to play its current role with its current status (Hulme & Edwards, 2013: 6–7). To this end, this study presents the following research questions: first, in the field of development, how did the role of CSOs change over time? Second, how has the government–civil society relationship undergone change? Third, what are the characteristics of the relationship during each period? To determine the answers to these questions, this study applies Najam's Four-Cs of Third Sector–Government Relations model, which presents a conceptual framework for the government–civil society relationship (Najam, 2000).

The structure of the paper is as follows. Section 2 presents a literature review and a conceptual framework for the analysis. In Sect. 3, changes in the role of CSOs are introduced in chronological order. Civil society history is divided into three periods: the first period is from 1995 to 2009, during which CSOs focused on service delivery. The second period is from 2010 to 2016, during which they played a role as an independent partner leading the major agenda of development cooperation and monitoring policies. The third period is after 2017, during which the foundation for a cooperative government–civil society relationship is being laid. The Four-Cs model of Najam is applied to derive the characteristics of the government–civil society relationship and to analyze this relationship's changes. Finally, Sect. 5 concludes the paper with suggestions for further study and a discussion regarding the direction of government–civil society partnership as well as its potential challenges.

2 Literature Review and Discussion on the Conceptual Framework for Analysis

The history of development CSOs has been discussed from various perspectives. Before the 1950s, voluntary relief activities were carried out, such as emergency relief. However, full-fledged international development cooperation activities on the part of CSOs are considered to have begun after the end of World War II (Bebbington, 2008; Clark, 1991). In the 1960s, wars of independence and conflicts increased within former colonial countries. Accordingly, CSOs actively carried out reconstruction activities and regional development activities, and the UN officialized the participation of CSOs (Sohn, 2015: 29). Through the 1970s and 1980s, the neoliberal approach centered on market-based reform, and the reduction of the state role was grafted into international development cooperation. The neoliberal theory recognized development CSOs as cost-effective private service providers (Kendall, 2003; Kim, 2017). Under the neoliberal theory, development CSOs went beyond simply delivering relief supplies and took the role of filling the gaps where the government's public spending and social services had been cut. Korten (1987) defines the expansion of the role of CSOs as a mobilizer period, or the "the second generation" for local development (Lewis & Kanji, 2009: 15). This period is also defined as the "early acceleration stage" by Bebbington and the "intensification" stage by Charnovitz (1997).

From the late 1980s to the 1990s, development CSOs faced an expansion period as they grew rapidly in terms of size and resources. By breaking away from excessive economism for development, a development discourse emphasizing human -rights-based approaches reappeared. The development discourse highlighted during this period presented alternative concepts such as participation, empowerment, gender, and people-centered approaches, and it advocated that development CSOs should recognize poverty reduction and development as issues of political rights and link them to the process of structural change. During this trend, CSOs in major donor countries became key development partners in emphasizing bottom-up processes and participating in policy decision-making (Lewis & Kanji, 2009: 47–70). McCormick (1993) argues that "the fundamental purpose of civil society is to influence public policy from outside the formal structure," and Smith and Lipsky (1993) explain that "the rise of public service contracting in the US strengthens this view of CSOs as policy entrepreneurs." Korten (1987) termed this period the "first generation" of CSOs, and Charnovitz (1997) referred to it as "empowerment." In the early 2000s, as the UN discussed the Millennium Development Goals (MDGs) for the eradication of poverty and the OECD DAC discussed the effectiveness of aid, development CSOs solidified their status as independent development actors and became a core development partner.

However, not all development CSOs have shifted their roles at the same rate. The relation depends on the state–social relationships within the donor country (Evers, 1995; Kim, 2008) and on the capabilities and identities of the CSOs. To illustrate this hiatus, this study aims to analyze the historical and social momentum through which development CSOs have grown and expanded their role and relationship with the government over the past 30 years, focusing on the Korean context. This approach contributes to a wholistic view of Korean civil society by breaking away from the dichotomous view of its roles as a program implementor and as a watchdog.

Adil Najam (2000) attempted a systematic analysis of government–civil society relations by applying a conceptual framework (the Four-Cs) based on the perspective of goals and preferred strategies for NGO–government relations. Najam's framework on the interaction between the two is highly useful in analyzing the changes in the strategic attitudes of the government and CSOs toward each other (Batley & Rose, 2011; Rurneaux & Ryan, 2014). The Four-Cs model has the advantage of making it easy to typify the role of developmental CSOs in relationships with governments. Another strength is that it can explain the change in the government–civil society relationship if the goals and preferred strategies of the two change over time.

Najam argues that although the government and CSOs clash in terms of goals, interests, priorities, and resources, the two can share strategic and institutional interests as CSO policy activities increase and the government expands the CSO policy-making process. Here, the relationship between the two is classified as cooperation, co-optation, complementarity, and confrontation based on whether or not the two have common goals and the same preferred strategies. Table 1 shows Najam's Four-Cs conceptual framework.

In particular, this model has the advantage of being able to explain the different strategic choices of each group, as it does not view the CSO sector or the government as a monolith. Despite the advantages described above, this framework model does not provide a sufficient explanation

Table 1 Four-Cs of NGO–government relations

		Goals (ends)	
		Similar	*Dissimilar*
Preferred strategies (means)	*Similar*	**Cooperation**	**Co-optation**
	Dissimilar	**Complementarity**	**Confrontation**

* *Source*: Najam (2000: 383)

for the dynamics of the government–civil society relationship. To provide an explanation of this relationship's dynamics, this study first examines in detail the changes in the role of development CSOs in each period in Sect. 3. This provides a background for explaining changes in the relationship between the government and CSOs.

3 Changes in the Role of CSOs in Development Cooperation: From a Service Delivery Implementor to a Partner

Starting with international relief activities in the 1990s and continuing into the 2000s, Korea's development CSOs actively responded to large-scale natural disasters, showing rapid growth in the number and scope of their activities. In terms of Korea NGO Council for Overseas Development Cooperation (KCOC) member organizations, 5, 34, 60, and 41 CSOs started international development projects before the 1990s, during the 1990s, during the 2000s, and during the 2010s, respectively (KCOC, 2017). CSOs are said to have been faithful in providing services in the 1990s and 2000s (Sohn, 2015: 56–68).

The early 2010s is pinpointed as the time period in which Korea emerged as an influential new donor in the field of development. The Korean government joined the DAC and hosted the G20 Summit and the OECD DAC HLF-4 in succession, taking the lead in global development discourse. Around that time, there were important changes in Korean development CSOs. The biggest change is that CSOs dealing with cross-cutting issues such as the environment, human rights, gender, and grassroots movements, who formerly focused on domestic issues, began conducting international development cooperation activities. As

CSOs began to speak out on related policies, the composition of development CSOs became diversified, and the scope of activities expanded (Kim, 2017; Lee, 2012; Sohn, 2015). In addition, experts in specialized fields such as medical health, education, water hygiene, and architecture began to be integrated into international development cooperation projects in various forms (KCOC, 2017). This change has become a bridge for the expansion of categories of international development projects and securing expertise while bringing diversity in methods of performance.

In addition, some development CSOs have begun advocacy and public awareness activities on development-related issues. Advocacy activities are carried out not only by the individual organization but also, more actively, by forming solidarity. The formation of the Global Call to Action Against Poverty (GCAP Korea) in 2008 as part of an international solidarity campaign to address poverty and inequality can be seen as the beginning. GCAP Korea is a policy and campaign network based on the voluntary participation of development CSOs. Member organizations, together with the United Nations, have held World Poverty Eradication Day every year on 17 October. It has also carried out advocacy activities including publishing a Korean-language version of the UN MDG report. GCAP Korea is significant in that Korea's development CSOs have formed the first network to connect with global civil society.

As international development cooperation was included as a main agenda item at the G20 Summit in Seoul in 2010, GCAP Korea held the Civil G20 Dialogue in cooperation with the G20 Preparation Committee, during which representatives of CSOs from around the world met with the G20 Sherpa. The Seoul Development Consensus for Shared Growth and the Development Agenda and Multi-year Action Plans, the outcomes of the G20 Summit, are the results of efforts on the part of development civil society and the Korean government. The outcome reports showed a positive attitude toward CSO participation. What is noteworthy is that in the process of preparing for the Civil G20 Dialogue, the development CSO network, which consists mainly of development CSOs, has served as an opportunity to expand contacts with Korean CSOs in other working areas. Development CSOs, which had been relatively passive in advocacy activities for the Korean government before the G20 Summit compared to CSOs in other fields, began conducting advocacy activities through civil society solidarity.

The following year, the 2011 OECD DAC Fourth High-Level Forum on Aid Effectiveness (HLF-4) was held in Busan. Based on the experience of the G20 Summit, CSOs formed the Korea Civil Society Forum on International Development Cooperation (KoFID) to serve as the Korean organizing committee for the Busan Civil Society Forum, a parallel meeting of the HLF-4. The KoFID played a role in gathering opinions from CSOs at the global level by participating in Better Aid, a global CSO network that led the issue of aid effectiveness, and the Open Forum, which led discussions on the development effectiveness of civil society since the Accra HLF-3 (Sohn, 2011).

KoFID activities have important implications in many regards. First, CSOs began to play a full-fledged role as watchdogs and policy partners. The KoFID declared that it aims to monitor Korean ODA policy as well as proposed aid and development policies based on universal values through strengthening the capabilities of Korean CSOs. In addition, the KCOC and the People's Solidarity for Participatory Democracy (PSPD) were co-chaired in forming the KoFID, establishing close communication and solidarity with CSOs in areas other than development. In addition, the participation of traditional CSOs, such as the Korea Human Rights Foundation, the Korea Green Foundation, the Korean Federation for Environmental Movements, and the Korea Women's Association United, has created balance among development CSOs and CSOs working on other agendas. Changes in composition affected the goals of the network, the way activities were carried out, and the scope of the agenda covered, thereby greatly expanding the scope of policy activities. After the HLF-4, the KoFID proposed policies by participating in international discussions on the government's policy establishment and implementation, including the OECD DAC peer review, the post-2015 development agenda, and the Open Working Group on Sustainable Development Goals. It also held forums, seminars, workshops, and meetings on international development cooperation policies, including the Seoul Civil Society Forum. At the global level, the KoFID has established networks and engaged in solidarity activities with the international community, such as the CSO Partnership for Development Effectiveness (CPDE) and the Asia Development Alliance (ADA).

Even before the establishment of the KoFID, there were individual organizations that worked to advocate for the development sector. ODA Watch, which has been active in advocacy since 2006, is a noteworthy example. ODA Watch was interested in engaging in development policy-making, but it did not have the structure and accountability as a representative of CSOs. The PSPD and the Citizens' Coalition for Economic Justice (CCEJ) also monitored the government's ODA and made efforts to intervene in related policies. The effort and interests of these individual organizations were the basis for the organization and activities of the KoFID as a development CSO network in the long term.

During the same period, there was also a change in development CSO field activities. CSO activities have expanded from service delivery to advocacy and public awareness. Traditional development cooperation sectors such as education, poverty, health, water, and sanitation account for the largest portion. However, in line with the trend of global development cooperation, such as linking with the UN SDGs and expanding humanitarian aid categories, agenda items such as jobs and economic growth, social economy, reduction of inequality, humanitarian assistance, migration and refugees, climate change, reduction of disaster risk, and psychological and social stability were discussed, and related projects began to be implemented (KCOC, 2017). Figure 1 below summarizes key events at the national and global levels as well as the CSO responses to them.

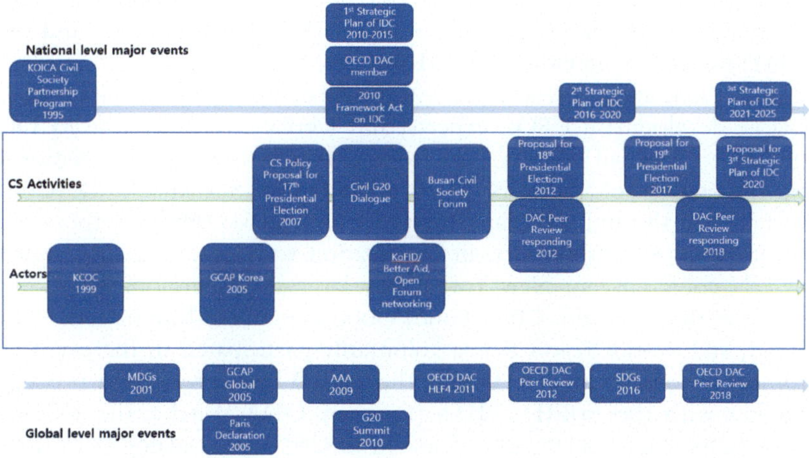

Fig. 1 Key events in development cooperation at the national and global levels and CSO responses (*Source* created by the authors)

On the role change of Korea's development CSOs, Lee (2012), Sohn (2015), and Kim (2017) argue that around 2010, CSOs expanded beyond their formerly limited role as an implementor to take on the role of an advocate by reinforcing the political advocacy function that criticizes and monitors the government at the stage of international development policy and project implementation. From the perspective of Charnovitz (1997) and Korten (1987), Korean development CSOs have followed the same path as global CSOs for 70 years, beginning in the 1950s, and in a compressed manner for about the past 30 years. According to the theory of Lewis and Kanji (2009), Korean CSOs have gone beyond the role of the first doer and the second catalyst and have taken a leap forward as a third-generation partner.

However, GCAP and the KoFID have inherent limitations in that they have been built up around major international conferences or global events. Both major CSO advocacy alliances were formed to respond to UN and OECD agendas, supported by a global network. GCAP focused on MDGs, and the KoFID focused on the OECD DAC HLF-4 on Aid Effectiveness. This has been the background for weakening the sustainability of each alliance since the process was terminated. Both networks were able to work very actively as the KCOC provided necessary resources, such as the secretariat and administration, while responding the global events. The advocacy capabilities and experiences, however, were not transferred to the policy function of the KCOC once GCAP and the KoFID weakened after the mid-2010s.

Since then, major advocacy activities have centered on a small number of large development CSOs, experts, and scholars. In 2019, the KCOC published the Civil Society Monitoring Framework for the National Implementation of the UN 2030 Agenda for Sustainable Development/SDGs through Human Rights and SDG16+, which is the SDG Interim Civil Society Report, in cooperation with experts and attended the UN conference in New York. Only large-sized development CSOs such as A Better World, Child Fund, Good Neighbors International, On Happy, and World Vision Korea voluntarily participated in the event to promote advocacy activities. During the OECD DAC peer review process, together with the KoFID, PIDA (formerly ODA Watch), the KCOC, and experts from academia participated in civil society dialogue to offer their opinions. Several CSOs such as Child Fund, Save the Children, and Good Neighbors International have actively carried out advocacy activities targeting the Korean government or through international conference channels hosted by UN agencies on specific issues such as the Convention on the Rights of the Child, refugees, and gender; these have produced fruitful results.

4 The Relationship between Government and Civil Society

Along with the growth of development CSOs, there have also been changes in the relationship and the level of cooperation between the government and CSOs. Government–civil society cooperation began in the field of ODA implementation. The first stage of the government–civil society partnership was launched in 1995. Cooperation between the Korea International Cooperation Agency (KOICA) and CSOs had long been maintained through very small-scale financial support until around 2010, when it expanded into a policy-making process. Section 4 analyzes the characteristics of the government–civil society relationship by dividing the period into three parts: the first period is from 1995 to 2009; the second one started from 2010, when the role of CSOs greatly changed; and the third is from 2017, when the government–civil society partnership began to be institutionalized.

4.1 First Period (1995–2009): Complementarity and Confrontation

Major donor countries have formed government–civil society partnerships and have carried out development projects based on them. Complementarity is a key principle of partnership in the process. For example, the government and CSOs share the common goals of international development cooperation, but at the same time, each has a distinct role to play. The following are examples of the contributions CSOs are determined to offer in international development cooperation that are different from those of the government (Han et al., 2019):

- Provide effective development service, especially for the poorest or the most vulnerable.
- Contribute to creating a democratic process in formulating and implementing development policies and strengthening the accountability of the government.
- Enable rapid response in the environment of humanitarian assistance due to disasters, disputes, etc.

- Claim and achieve citizens' rights for good governance in developing countries.
- Expand the space for local people to speak out in policy dialogue in developing countries.

Thus, in general, the government seeks to effectively achieve the common goals of international development cooperation by maximizing the characteristics and strengths of the CSOs by creating a channel to support them. KOICA's CSO support program is also based on this principle of complementarity. The program focuses on financial support for CSOs. In the early stage, KOICA supported CSOs through the Matching Grant Program, providing subsidies within 50% of the total budget of the projects. In 1995, KOICA's financial support for CSOs amounted to KRW 489 million, which accounted for around 1.2% of the total ODA. Since then, KOICA has applied for 1,945 projects over the course of 24 years under the CSO support program. The total budget for the project has been reported as approximately KRW 32.8 trillion. The steady growth in the scale of CSO support means that many CSOs have applied to the programs and implemented Korean ODA. This can be interpreted as CSOs agreeing to government goal. It is said that a complementary relationship between the two was formed in terms of the implementation of aid and continued afterward.

At the same time, it is noteworthy that from 1995 to the late 2000s, government–civil society cooperation was limited to the area of service implementation and focused on financial support for CSOs. CSO participation in the decision-making process was very limited because there was no institutional channel for CSOs to express opinions on the development cooperation agenda on a policy level. In other words, while the government and CSOs cooperated for the provision of aid services, there was no interaction during policy-making. CSOs presented their opinions by issuing statements or proposals when there were important development cooperation issues. The CSO statement on the Framework Act on Development Cooperation in 2009 is a good example. In the statement, CSOs pointed out the issue of ODA fragmentation and opposed loans and tied aid. It can be reasonably inferred that the CSOs recognized that ODA implementation methods, such as loans and tied aid, do not correspond to the goals of international development cooperation because Korea's economic interests might be considered first. The CSO statement on the Framework Act on Development Cooperation also showed CSOs' confrontation with the government. Moreover, the fact that CSOs did not have the opportunity and space to express their opinions shows that the government also did not consider CSOs as partners to achieve the goals of development cooperation together.

4.2 Second Period (2010–2016): A Mixture of Complementarity, Confrontation, and Cooperation

The Korean government, as an emerging donor, has declared its compliance with international development norms and has actively embraced the international community's development discourse centered on the UN and the OECD DAC. Since 2010, the development discourse has focused on expanding the role of CSOs as independent development actors and on building a horizontal relationship between the government and CSOs. This is due to the fact that the UN SDGs are based on a bottom-up process and the OECD DAC, which emphasizes the participation of various actors in development, especially CSOs, has led the discourse. The Korean government joined the international development community as a member of the OECD DAC and CSOs as members of the transnational CSO network. In the process, Korea has, in fact, accepted the international norms discussed at the global level (Lee, 2012; Sohn, 2015). Against this background, in the process of hosting the G20 Summit and the HFL-4, the government increased formal and informal policy talks with CSOs. In other words, the expansion of the channel of government–civil society cooperation in Korea is strongly linked to the trend of emphasizing government–civil society partnership in the international community. As a result, this initiated a change in the government's attitude toward CSOs.

At the same time, CSOs began full-fledged advocacy activities, forming advocacy networks for policy intervention such as KoFID, as described in Section 3. Table 2 below shows the CSO policy networks and their major activities. It also presents the fact that CSO advocacy activities more than doubled between 2010 and 2016 compared to those during the 2005–2009 period. During the second period, CSOs criticized the government for downplaying efforts to achieve the MDGs and tried to force government efforts, issuing a statement urging the achievement of the MDGs in 2010. CSOs also engaged in confrontation with the government by providing an opinion on the Enforcement Decree of the Framework Act on International Development Cooperation in 2013. They request

Table 2 CSO policy networks and their major activities

Year	*Issues/events*	*Actors*	*Means*
2005	GCAP Korea	KCOC, CCEJ, YMCA	GCAP Korea launched
2007	Response to the 17th presidential election	KCOC, GCAP Korea	Proposal
2009	Framework Act on Development Cooperation	GCAP Korea, PSPD, YMCA, ODA Watch	Statement
2010	MDGs	ODA Watch	Submit a written opinion
2010	Framework Act on Development Cooperation/ Enforcement Decree	PSPD	Submit a written opinion
2010	G20 Summit	GCAP Korea	The G20 Civil Dialogue
2010	KoFID	GCAP Korea, KCOC, PSPD	KoFID launched
2011	OECD DAC HLF-4	KoFID	Busan Civil Society Forum
2012	2012 DAC peer review	KoFID, KCOC, ODA Watch, CSOs, scholars and experts	Dialogue, submit a written opinion
2012	Response to the 18th presidential election	KoFID, KCOC, KAIDEC, GCAP	Proposal, discussion meetings, and open inquiries for presidential candidates
2013	The Enforcement Decree of the Framework Act on International Development Cooperation	KoFID	
2014	Korean government's plans to join IATI	KoFID, PSPD	Welcome comments
2015	The Second Strategic Plan–Mid-term Strategy	KoFID	Statement
2015	SDGs	KoFID	Submit a written opinion
2017	Committee for International Development Cooperation (CIDC)	KoFID	Submit a written opinion

(continued)

Table 2 (continued)

Year	*Issues/events*	*Actors*	*Means*
2017	Response to the 19th presidential election	KoFID	Proposal, discussion meetings, and open inquiries for presidential candidates
2018	2018 DAC peer review	KoFID/KCOC- DAC Peer Review CS FT	Publishing reports, dialogue
2020	The Third Strategic Plan–Mid-term Strategy	KoFID/KCOC, PSPD	Proposal, public hearing, and dialogue

* *Source* created by the authors by referring to web pages and white papers on the network

a full review of the act, claiming that it had not improved the structural problems of Korean ODA.

CSOs, however, did not just criticize the government. Together with GCAP Korea and a few advocacy CSOs, KoFID requested to be recognized as a partner to the government and demanded political opportunities and space from both the National Assembly and the government. KoFID, KCOC, and some other CSOs were officially invited to participate in providing suggestions during the OECD DAC peer review process. Moreover, CSOs presented proposals to the 18th presidential candidate campaigns and transition committees. KoFID and PSPD welcomed the Korean government's plans to join the International Aid Transparency Initiative (IATI) in 2014 and announced their support for the government's ODA goals and strategic means in a statement of opinion regarding the Committee for International Development Cooperation (CIDC) in 2017 and the 2018 DAC peer review. This stance shows the possibility of mid- to long-term cooperation.

Summarizing the activities of these CSOs, the government–civil society relations during this period were very complex. The relationship between the two, as shown during this period, had all the characteristics of complementarity, confrontation, and cooperation. Along with government–civil society relations, it should be noted that the role of civil society has expanded from a simple implementor to policy advocator.

Nevertheless, it is difficult to say that the change in the role of CSOs is due to changes in the overall activities of the majority of them. Rather, it should be noted that these changes were heavily dependent on the activities of several large CSOs as well as some expert active in advocacy activities. GCAP and the KoFID, which were the pillars of development cooperation advocacy activities around 2010, consisted of 35 and 22 member organizations, respectively. The total number of CSOs who actively engaged in policy activities was 43, excluding overlapping organizations. This accounted for only 37% of the total 116 CSOs as of 2013. Both GCAP and the KoFID worked together with a relatively diverse range of actors through cooperation with networks of CSOs working in other fields. As a result, 50 Korean CSO representatives and 150 global CSO representatives participated in the parallel meetings of the 2010 G20 and OECD DAC HLF-4, respectively. However, once again, it is difficult to declare that such achievements represent a change in the members of the development CSOs overall.

On the other hand, the implementation of service delivery for development projects carried out by government–civil society partnerships has not changed much since 2010. During the projects, there are often complaints that the government still recognizes CSOs as a mere facilitator rather than a partner. It has been pointed out that this is the result of the institutionalization of administrative opportunism and the government's recognition of CSOs as a facilitator to provide development services (Kim, 2017: 131). In other words, an unbalanced complementary relationship is maintained between the government and CSOs. Therefore, it is difficult to conclude that the partnership has been strengthened to the same level in all aspects of development cooperation. Taken together, the asymmetrical complementary relationship and the confrontation continue. At the same time, cooperation also exists.

4.3 *Third Period (After 2017): Laying the Foundation for a Cooperative Relationship*

The expansion of CSO participation in ODA policy-making and some institutionalized partnership governance are key to government–civil society partnerships. Until 2017, there were no overarching policy documents containing comprehensive partnership agreements between governments and CSOs. There were only sub-strategic documents such as the Customized Support Plan for CSOs in the Development Cooperation Field, which was announced in 2016, and KOICA's strategy as part of implementing its plan. Moreover, these documents are unilaterally prepared by the government. CSOs have continuously suggested that

CSOs need an overarching government–civil society partnership strategy to ensure the stable participation of CSOs in policy decisions.

Around 2017, the institutionalization of CSO cooperation began in response to the continued demands of CSOs. Experts and representatives of CSOs became members of the CIDC and its Working Committee at the Office for Government Policy Coordination. These committees became channels for CSOs to add their opinions to the development-related policy-making process. KOICA, an aid agency, also includes experts as board members. The efforts of both sides for policy cooperation finally led to the establishment of policies that will serve as the basis for government–civil society cooperation in the future development field as of 2019.

With the establishment of the Policy Framework for Government–Civil Society Partnership in International Development Cooperation in 2019, the government established an institutional framework and foundation for participation in development CSOs in terms of policy. The OECD DAC peer review influenced the establishment of the policy framework for partnership. The OECD DAC peer review conducted in 2017 advised that "the Korean government needs to clarify and deepen cooperative relations with CSOs by establishing a normative framework that recognizes the various roles of CSOs as an implementation partner and independent development cooperation actor" (DAC, 2017). In 2018, the OECD DAC recommended that the Ministry of Foreign Affairs prepares the Government–Civil Society Partnership Framework in the Field of International Development Cooperation. The Ministry of Foreign Affairs, which was in charge of this, proposed a joint creation to CSOs. The policy framework for partnership was passed at the 32nd CIDC held in January 2019 through a consultation process between the government and CSOs.

In 2020, KoFID, together with KCOC and PSPD, strongly criticized the government's Third Strategic Plan–Mid-term Strategy for leaving room to pursue national interests. Considering the opinions, it can be seen that there is still a large gap between the two. However, in the process of establishing the government's strategy, the foundation for cooperation was laid in that opportunities to hear the voice of civil society were systematically established through public hearings and civil society dialogues.

5 Conclusion

In short, in observing the growth path, it can be found that a series of key international conferences held in Korea served as important momentum for development CSOs to grow into watchdogs and partners in the policy-making process. Korean development CSOs have fully utilized major international conferences held in Korea as political opportunities and space. Despite the short period to prepare, Korean CSOs actively formed an alliance in conducting advocacy activities and in successfully holding parallel meetings responding to each conference. Through this process, development civil society, which was composed only of development CSOs working on international development cooperation until the 2010s, became a full representative of a civil society through solidarity with other fields. At the same time, Korean CSOs secured their expertise by combining scholars in the field of international development cooperation and sector experts in the fields of human rights, gender, and environment. These experiences have brought significant changes to the role of CSOs, which have traditionally been recognized as service delivery implementors with a focus on project implementation, in the development sector.

However, this is mainly the achievement of a small number of large CSOs along with exceptional scholars and experts. Therefore, development CSOs have not accumulated the common experience of changing the political and social environment on their own. In this process, the issues of accountability and representativeness still remain, as scholars, experts, and CSO representatives participating in the official channel have not secured or been given representation to collect opinions from all development CSOs. Additionally, the overall level of participation in CSOs is still low because institutionalized CSO participation in actual policy-making process is quite limited.

To summarize, Korean CSOs were able to utilize external factors and environmental opportunities to promote internal growth. According to the results of analyzing the government–civil society relationship by applying Najam's framework, the government–civil society relationship changed from a complementary/confrontational relationship in 1995–2010 to a mixture of complementary/confrontation/cooperation in 2011–2016, and the foundation for a cooperative relationship laid after 2017. The change in the role of CSOs and the Korean government's acceptance of international norms have been crucial to the change in the government–civil society relationship. However, while development CSOs as a whole can be seen to have taken a common path of development that appears in the international trend, it is difficult to judge whether these changes in activities and roles are common among development CSOs overall. To fill this gap, Kim (2017) states that Korean CSOs need historical experience in leading a positive policy and system transformation for the government.

In this context, the policy framework for partnership could serve as an institutional basis to facilitate the historical experience of development CSOs in the future. In terms of content, the policy framework for partnership is largely composed of partnership goals and objectives, partnership principles, and partnership implementation plans. This is important in that KOICA has defined the purpose, goals, and principles of government–civil society partnership in the field and proposed ways to implement the partnership for the first time in the 24 years since it started supporting CSO projects in 1995. It is also meaningful in that it is a government policy document prepared by the government and CSOs jointly participating in and discussing its formation process. Although there are limitations in content and formality, it is expected that it will be used as a framework for establishing the direction of policies, strategies, and partnerships among government ministries and affiliated organizations, public aid agencies, and development CSOs. The trend of strengthening policy cooperation that has emerged since 2017 suggests a direction for the government and CSOs to move toward cooperation. If this is institutionalized, cooperation is expected to be strengthened.

References

Batley, R., & Rose, P. (2011). Analysing collaboration between non-government service providers and governments. *Public Administration and Development, 31*(4), 579–585.

Bebbington, A., Hickey, S., & Mitlin, D. (2008). *Can NGOs make a difference? The challenge of development alternatives*. Zed Books.

Clark, J. (1991). *Democratizing development: The role of voluntary organizations*. Kumarian Press.

Charnovitz, S. (1997). Two centuries of participation: NGOs and international governance. *Michigan Journal of International Law, 18*(2), 184–286.

Furneaux, C., & Ryan, N. (2014). Modelling NPO–Government relations: Australian case studies. *Public Management Review, 16*(8), 1112–1140.

Han, J., Sohn, H., Kim, S., Shin, J., Kim, M., Lee, H., & Rha, S. (2019). *Study on implementation plan for policy framework for government–Civil society partnership in international development cooperation*. KOICA.

Hulme, D. & Edwards M. (2013). *NGOs, states, and donors too close for comfort?* IPE classics. 2nd ed. Palgrave Macmillan.

KCOC. (2017). *International development cooperation csos statistics handbook*. http://www.ngokcoc.or.kr/bbs/board.php?bo_table=paper01&wr_id=152. (last access: 17th August, 2020).

Kendall, J. (2003). *The Voluntary sector*. Routledge.

Kim, E., & Hong, J. (2014). A significant role of global civil society in forming international development cooperation norm: focusing on OECD DAC high level forum on aid effectiveness. *Journal of International Politics, 19*(1), 175–206.

Kim, T. (2015). The fragmentation of global governance for development cooperation in the post-2015 development era: The UNDCF and the GPEDC compared. *Journal of International Area Studies, 24*(3), 95–126.

Kim, T. (2017). From supporters to advocates: A study on the identity transition of development civil society organizations in South Korea. *Discourse 201*, 20(3), 111–155.

Korten, D. (1987). Third generation NGO strategies: A key to people-centered development. *World Development, 15*(1), 145–159.

Lee, S. (2012). The role of CSOs in international development cooperation: After HLF-4 2011. *Journal of International Development Cooperation, 7*(1), 89–115.

Lewis, D., & Kanji, N. (2009). *NGO and development*. Routledge.

McCormick, J. (1993). International nongovernmental organizations: Prospects for a global environmental movement. In S. Kamieniecki (Ed.), *Environmental politics in the international Arena: Movements, parties, organizations, and policy*. State University of New York Press.

Najam, A. (2000). The four-C's of third sector-government relations cooperation, confrontation, complementarity, and co-optation. *Nonprofit Management & Leadership, 10*(4), 375–396.

OECD DAC. (2018). *OECD development co-operation peer reviews: Korea 2018*. OECD. https://www.oecd.org/korea/oecd-development-co-operation-peer-reviews-korea-20189789264288829-en.htm(last access: 11th September, 2020).

Smith, S. R., & Lipsky, M. (1993). *Nonprofits for hire: The welfare state in the age of contracting*. Harvard University Press.

Sohn, H. (2011). Busan HLF-4 and global civil society response strategies: Focusing on better aid and open forum. *Journal of International Development Cooperation, 6*(3), 115–131.

Sohn, H. (2015). *Civil society and international development cooperation*. JipMoonDang.

Sohn, H., Han, J., & Park, B. (2011). A study of government-civil society partnership in international development cooperation: Focusing on the NGO assistance policies of OECD DAC members. *National Strategy, 17*(4), 105–136.

CHAPTER 11

Japan's Approach to the SDGs: Decoupling Between the SDGs and International Development

Tatsufumi Yamagata

1 Introduction

The Sustainable Development Goals (SDGs) were created as an offspring of the Millennium Development Goals (MDGs) in September 2015. While the MDGs are those for the sake of international development, the SDGs cover more than that. The SDGs accommodate economic and social development in developed countries, as well as the environmental conservation of the world. Some say that the SDGs serve the people who were "left behind" in the domain of the MDGs. These were people in developed countries and the future generations of the world.

This is in accordance with Japan's shift in attitude toward international development. In February 2015 the Cabinet of Prime Minister Shinzo

T. Yamagata (✉)
Ritsumeikan Asia Pacific University, Beppu, Japan
e-mail: yama-apu@apu.ac.jp

H.-j. Kwon et al. (eds.), *International Development Cooperation of Japan and South Korea*,
https://doi.org/10.1007/978-981-16-4601-0_11

Abe approved the Development Cooperation Charter, which replaced the ODA Charter. The ODA Charter, which was adopted in 1992, had been the backbone of Japan's international cooperation. An important difference between the two charters is that the Development Cooperation Charter explicitly refers to Japan's national interests, which are expected to benefit from international cooperation, while the ODA Charter did not. Thus, the feature of the SDGs to benefit a country's own nationals coincides with the manifestation of the pursuit of national interests in the Development Cooperation Charter. For the Government of Japan, replacement of the MDGs with the SDGs was helpful to justify the conversion of the character of international cooperation towards the aim of national interests under the new label of "development cooperation."

This chapter argues that Japan's efforts to achieve the SDGs are mainly directed toward Japanese nationals to convince them of the meaningfulness of the efforts and that international development to benefit people in need in developing countries is sidelined. This author calls this bifurcation "decoupling between the SDGs and international development." The concept of sustainable development in line with the SDGs is more frequently recognized in the Japanese development of local economies in Japan and ecofriendly development in the rest of the world. Development and human rights issues in developing countries are less likely to be recognized by the Japanese people because they are remote and therefore of little concern. This tendency has been accentuated under the COVID-19 pandemic in 2020, which makes people unable to travel abroad and enforces in-house and online activities for almost everyone in the world.

The rest of this chapter is organized as follows. In Sect. 2, Japan's involvement in the processes of formation of the SDGs from 2012 through 2015 is described. Japan's involvement was reactive to actions to form the 17 goals and 169 targets of the SDGs. In Sect. 3, ongoing institutional frameworks to advance the SDGs are analyzed. Generally speaking, the frameworks are formed to appeal to the Japanese for participation. There are some duplicities of policies written in Japanese and English such that the former highlights effects on the Japanese, while the latter weakens the focus on the Japanese. Section 4 describes the involvement of the private sector in the promotion of the SDGs in Japan. Japan is a country where the SDGs maintain momentum through the public, while climate change and COVID-19 overrode the SDGs in many other countries as of 2020. A reason for this popularity of the SDGs in Japan lies in the engagement of the private sector. The final section concludes this chapter.

2 Japan's Participation in the Formation of the SDGs

In the first decade of this millennium, the incidence of poverty was remarkably reduced in many parts of the world in economic and social terms. The Millennium Development Goals, which were effective from 2000 through 2015, were synchronized with this achievement. Therefore, a similar framework of goal-setting, which takes over where the MDGs left off, was considered sensible. Before the SDGs materialized, this issue was called the "post-2015 development agenda."

The post-2015 development agenda was finally renamed the Sustainable Development Goals and released at the United Nations Conference on Sustainable Development, which took place in Rio de Janeiro, Brazil, in June 2012. The conference was also called Rio+20. At the conference the adoption of the SDGs as the successor of the MDGs was settled.

This section indicates how Japan struggled to keep up with the agenda-setting processes to formulate the SDGs and their targets.

2.1 Japan's Post-2015 Development Agenda: Pact for Global Well-Being

Three months before Rio+20, the United Nations Development Programme (UNDP), which served as a coordinator to promote the post-2015 development agenda on behalf of the United Nations, held a consultation meeting for the East Asian region.[1] In the meeting resource persons from East Asian countries, including former ministers of finance, were consulted about prospective directions of the post-2015 development agenda (Yamagata, 2016). A main issue of consultation was whether the post-2015 global goals should be "development-focused," as the MDGs were, or "universal" as beneficiaries of the new global goals should contain a wider range of people than those in developing countries (UNDP, 2012). It is obvious that the universality principle accords

[1] The title of the meeting was "East Asian Regional Consultation on Human Development Report and Measurement of Human Development Progress," held on 13 March 2012 at the JICA Research Institute.

with the SDGs, which were released three months later at Rio+20. Thus, the UNDP sounded out the resource persons from East Asia about the universality feature of the SDGs at this consultation meeting. On this occasion, a delegate from the Global Issues Cooperation Division of the Ministry of Foreign Affairs of Japan proposed a new concept for the post-2015 global goals, entitled the Pact for Global Well-being. This concept consisted of (1) a new guiding principle entitled "well-being for all," (2) a new set of goals for "measuring multifaceted results," and (3) a new set of enabling conditions composed of green growth, inclusive growth, shared growth, and knowledge-based growth. The concept of "well-being" included human security, sustainability, equity, and mutual support. Human security is Japan's signature principle for pursuing peaceful and economically stable lives (Commission on Human Security, 2003; Kamidohzono et al., 2016). It is noticeable that the division of the Ministry of Foreign Affairs (MOFA), which is one of the best-informed sections of the Government of Japan concerning the post-2015 agenda, proposed the new concept for the post-2015 global goals without knowing about the release of the SDGs at Rio+20 three months later. Even MOFA Japan was not well-informed about the progress in determining the post-2015 development agenda.

2.2 *Open Working Group on Sustainable Development Goals*

The Government of Japan had difficulties influencing the contents of the SDGs. In order to elaborate details of the SDGs and their targets, the United Nations General Assembly decided to establish the Open Working Group of the General Assembly on the Sustainable Development Goals in January 2013 (UNGA, 2013). The Open Working Group (OWP) is composed of 30 subgroups of countries. A subgroup consists of one to four countries. The subgroup is a unit of countries by which the goals and targets of the SDGs are proposed. Each UN member country freely negotiates with one another and formulates a subgroup. Like-minded countries tend to form a subgroup because these countries can reach agreement on the SDG goal and targets proposals easily. Examples of subgroups are the following: Algeria, Egypt, Morocco, and Tunisia; Zambia and Zimbabwe; Nauru, Palau, and Papua New Guinea; Bhutan, Thailand, and Vietnam; India, Pakistan, and Sri Lanka; China, Indonesia, and Kazakhstan; Australia, the Netherlands, and the UK; and Canada, Israel, and the United States. Japan formed a subgroup with Iran and

Nepal, while the Republic of Korea made a subgroup with Bangladesh and Saudi Arabia (UNGA, 2013). Though it was recommended that these subgroups would be units of goals and targets for proposal, proposals made by a single country were also allowed in the end.

The OWG meeting was held 13 times from March 2013 through July 2014. The frequency of the meeting was almost once a month (13 times for 17 months). The role of the OWG was to prepare a long list of goals and targets for the prospective SDGs and to narrow them down into a short list. A long list was compiled after the 10th OWG meeting in March 2014 and publicized as Encyclopedia Groupinica (OWG-SDGs, 2014). There remained 19 goals and 1994 targets, which were reduced to 17 goals and 169 targets in OWG's final report submitted to the UN General Assembly in August 2014 (UNGA, 2014). Thus, one year and one month before the SDGs were approved at the UN General Assembly in September 2015, the overall structure of the SDGs was decided.

From the first meeting in March 2013 through the tenth meeting in March 2014, the preformed country subgroups, individual UN member countries, some groups of member countries, and some NGOs proposed various ideas for the goals and targets of the SDGs. Since this set of new international goals was expected to influence the direction of global socioeconomic development, all were keen to put forward their own agendas as much as possible. An interim report on the proposals was published in Encyclopedia Groupinica. Therefore, as many as 1994 proposals for targets were submitted.

Table 1 exhibits how actively Japan and other representative countries were involved in forming the contents of the SDGs. Subgroup-wise submission was the norm of the OWG. France submitted 89 proposals jointly with Germany and Switzerland. The UK submitted 84 target proposals with Australia and the Netherlands. Likewise, the United States submitted 64 proposals with Canada and Israel, while the United States had another proposal submitted as a single country. China made 31 target proposals with Indonesia and Kazakhstan, while the country proposed 10 more targets in collaboration with the Group of 77. In total, China was involved in 41 submissions all together.

From Table 1, it is obvious that Japan's contribution to the SDG target design was small. The country submitted only three proposals. The three target proposals were as follows: (1) "by 2030, at least 80% of the poorest 40% of the population have coverage to ensure access to essential health services" (footnote #350 of Encyclopedia Groupinica);

Table 1 Number of targets suggested for the SDGs in *Encyclopedia Groupinica*

	Japan	*Republic of Korea*	*China*	*US*	*UK*	*France*
Subgroup entries	0	2	31	64	84	89
Single-country entries	3	13	0	1	0	0
Other groups	0	0	10	0	0	0
All entries	3	15	41	65	84	89

Source OWG-SDGs (2014)
Note The total number of suggested targets is 1994. Subgroups of the countries indicated above are as follows: Iran, Japan, and Nepal; Bangladesh, the Republic of Korea, and Saudi Arabia; China, Indonesia, and Kazakhstan; Canada, Israel, and the United States; Australia, the Netherlands, and the UK; and France, Germany, and Switzerland. One of the "other groups" for China includes the Group of 77 and China. The Group of 77 is a coalition of developing countries, which was initiated in 1964 with 77 founding member countries at the United Nations Conference on Trade and Development (UNCTAD)

(2) "achieve Universal Health Coverage, where all people have access to the quality, essential health services they need without enduring financial hardship" (footnote #364); and (3) "by 2030, everyone has coverage to protect them from financial risk, so that no one is pushed into poverty or kept in poverty because of expenditure on health services" (footnote #365). All three suggestions are located in Focus Area 3. Health and Population Dynamics. It is notable that all three concern universal health coverage (UHC),[2] which is a signature policy of Japan's international cooperation. UHC is particularly mentioned in the Development Cooperation Charter (Cabinet, 2015) and Japan's voluntary national review for the SDGs (Government of Japan, 2017).

At a glance, the number of target proposals submitted by Japan is extremely low in comparison with the other developed countries exhibited in Table 1. Another observation on Japan's proposals is that there was no proposal jointly submitted as a subgroup with Iran and Nepal. No subgroup submission of target proposal seems to imply that the three-country collaboration did not work well. Iran made 13 proposals for SDG target as a single country, which is more than the number Japan

[2] This is used interchangeably with universal healthcare. Both are abbreviated as UHC. UHC is pursued by the World Health Organization, too. It was taken as the main theme of the *World Health Report* in 2013 (WHO, 2013).

made.[3] Among 13 proposals there was a target concerning UHC, which is associated with footnote #381. However, Iran's proposal on UHC was submitted independently of Japan's three UHC-related proposals associated with footnotes #350, #364, and #365. As a result, Japan's influence on goal and target setting was very limited. Numerically, three proposals out of a total of 1994 amount to 0.15%, which are substantially inferior to the same ratios for France, the UK, the United States, China, and the Republic Koreas of 4.5%, 4.2%, 3.3%, 2.1%, and 0.75%, respectively. It seems that France, the UK, and the United States formulated subgroups with countries that keep regular and frequent communication, so collaboration for a very short period between March 2013 and March 2014 may have been smoother for their subgroups than that of Iran, Japan, and Nepal.

Japan was proud of its contribution to the formation of the international development goals at the OECD/DAC, which originated the Millennium Development Goals (JICA ed., 1998; MOFA, 1997; OECD/DAC, 1996a). As for the SDGs, however, Japan was not deeply involved in the SDG formation process. The release of news on the replacement of the MDGs with the SDGs at Rio+20 in June 2012 was not notified to the responsible ministry in Japan three months before. After beginning the elaboration process of the contents of the goals and targets of the SDGs, Japan did not catch up with other countries, which quickly negotiated with like-minded partner countries and produced their agendas effectively as targets of the SDGs. Generally speaking, Japan was reactive to the formation of the SDGs and did not influence the contents of the new global goals very much.

3 Institutional Framework to Advance the SDGs

This section discusses main issues regarding the promotion of the SDGs in Japan after they came into effect in September 2015. Meanwhile, there was an important policy change taking place a half-year before that in Japan. That was the replacement of the Official Development Assistance (ODA) Charter with the Development Cooperation Charter. The new

[3] Nepal did not make any target proposal as a single country. The country belongs to the groups "landlocked developing countries" (LLDC) and the Group of 77. LLDC and the Group of 77 plus China submitted some target proposals as groups.

charter symbolizes the "inward-looking" orientation of Japan's international cooperation, which is applicable to the promotion of the SDGs in Japan, too.

3.1 The Development Cooperation Charter and the SDGs

3.1.1 Manifestation of National Interests in the Development Cooperation Charter

The ODA Charter had been the backbone of Japan's international cooperation since it was adopted in 1992. The original ODA Charter declared that Japan's ODA would pursue both "environmental conservation and development".[4] Then, a revised ODA Charter in 2003 spelled out its objectives as follows:

> The objectives of Japan's ODA are to contribute to the peace and development of the international community, and thereby to help ensure Japan's own security and prosperity. (Section I. Philosophy: Objectives, Policies, and Priorities of MOFA, 2003)

Thus, Japan's ODA pursues "the peace and development of the international community" and as fruits of this endeavor accrues "security and prosperity" for Japan. This was the main logic to link the benefits of Japan's ODA to the international community with indirect benefits for Japanese nationals. Even while the original and revised ODA Charters were in force in the 1990s and the 2000s, some scholars indicated the tendency of Japan's ODA to be more likely granted by Japanese firms or their joint ventures (Arase, 1995; Ensign, 1992; Leheny & Warren, 2010; Orr, 1990). OECD/DAC was also critical of a high ratio of "tied projects," which require procurements from Japanese companies, and kept monitoring this ratio, which declined to zero as long as ODA loans were concerned at the end of the 1990s (OECD/DAC, 1996b, 1999).

Such efforts to wipe out the criticism of Japan's ODA regarding favoritism for Japanese firms were abandoned in the Development Cooperation Charter. The Development Cooperation Charter was adopted by the Cabinet of the Government of Japan in February 2015. This new

[4] See the following site for an English translation of the original ODA Charter: https://www.mofa.go.jp/region/n-america/us/q&a/oda/3.html.

charter spelled out "national interests" as an aim of "development cooperation." This new concept of development cooperation is broader than that of ODA in terms of two dimensions, as shown in Fig. 1.

The first dimension represents contributors. While ODA was solely owed by the central government, development cooperation is shouldered by the private sector, Japanese local governments, and NGOs. The second dimension represents recipients. While recipients of ODA were limited to central governments of developing countries, recipients of development cooperation are expanded to include middle- high-income countries with "special vulnerabilities" such as natural disasters, infectious diseases, and climate change issues. For example, assistance with a pure water supply to high-income Middle Eastern countries in a dry climate may be counted as development cooperation. In addition, Japanese small- and medium-sized enterprises (SMEs) are also included as recipients of development cooperation if an SME serves a developing country in a designated program such as a base of the pyramid (BOP) scheme, which promotes Japanese SMEs conducting business for the sake of the lowest-income group of people in developing countries (Yamagata, 2016).

Finally, the Development Cooperation Charter specially mentions a "proactive contribution to peace" through development cooperation even though it is emphasized that this is done for non-military purposes. An

		Contributors	
		Government of Japan	Private sector/local governments/NGOs
Recipients	Developing countries	ODA	Development cooperation
	Middle- and high-income countries with vulnerabilities	Development cooperation	
	Japanese SMEs		

Fig. 1 Recipients of and contributors to development cooperation (*Note* This is a modified figure that appears in Yamagata (2016). ODA is a part of development cooperation, which includes resource transfers from the private sector, local governments, and NGOs as well as those directed to middle- and high-income countries with vulnerabilities and Japanese SMEs. The domain of development cooperation is encompassed by the broken line.)

example of cooperation in this context is assistance facilitated through a national coast guard.

3.1.2 Common Features of the SDGs and the Development Cooperation Charter

What deserves special attention is the fact that the features of the Development Cooperation Charter mentioned in the previous section are common in the SDGs. They include (1) the expansion of areas of cooperation toward non-development fields, (2) expansion of the domain of beneficiaries toward non-poor people, and (3) burden sharing to the private sector and NGOs. Let us discuss these common features in order.

Table 2 displays SGD 17 goals. The MDGs have eight goals that were about poverty, education, gender equality, children's health, maternal health, infectious diseases, environmental sustainability, and global partnership. Roughly speaking, the eight goals correspond to Goals 1, 3, 4, 5, 6, and 17 of the SDGs. Goals 11, 12, 13, 14, and 15 of the SDGs are new additions to the MDGs to elaborate factors of environmental sustainability. Goals 7, 8, and 9 may be interpreted as a means to achieving the ultimate goal of people's well-being. Goals 10 and 16 are also new additions to address inequality, peace, justice, and governance.

Thus, the domain of achievements of the SDGs is richer and more diverse than that of the MDGs. This aspect may be a consequence of

Table 2 Sustainable development goals

No	*Goal*	*No*	*Goal*
1	No poverty	10	Reduced inequality
2	Zero hunger	11	Sustainable cities and communities
3	Global health and well-being	12	Responsible consumption and production
4	Quality education		
5	Gender equality	13	Climate change
6	Clean water and sanitation	14	Life below water
7	Affordable and clean energy	15	Life on land
8	Decent work and economic growth	16	Peace, justice, and strong institutions
9	Industry, innovation, and infrastructure	17	Partnership for the goals

Source the website of the SDGs from the Department of Economic and Social Affairs, United Nations (https://sdgs.un.org/goals)

great progress in poverty reduction in many parts of the world during the time covered by the MDGs. Hence, the international community wanted the SDGs to have a wider range of issues with tougher targets than those of the MDGs.

The switch from the ODA Charter to the Development Cooperation Charter in Japan also expanded the domain of targets for the Japanese. While the ODA Charter exclusively dealt with development in developing countries, the Development Cooperation Charter extended its scope to the peace and security of Japan and its neighbors, even though the non-military nature of this new orientation is maintained.

A second common aspect between the SDGs and the Development Cooperation Charter is the expansion of beneficiaries beyond people in developing countries. As for the SDGs, environmental sustainability is supposed to benefit everybody on the globe. On top of that, the SDGs have a universality principle, which is often translated into the slogan "leave no one behind". This could be interpreted as incorporating deep coverage of the poorest and the most underprivileged people. However, some others interpret this slogan as meaning that people in rich countries should benefit from the SDGs.[5] Indeed, for a voluntary national review of the SDGs, even developed countries examine the development of their own economies and societies first, then review international cooperation with developing countries subsequently. The performance measures in terms of progress toward the SDGs are mainly those of achievements in domestic economies and societies.[6] As detailed in the previous section, the Development Cooperation Charter highlights Japan's national interests more than its predecessor, the ODA Charter.

Thirdly, burden sharing from the public sector to the private sector is a common and outstanding feature of both the Development Cooperation Charter and the SDGs. Figure 1 evidently shows that private companies, philanthropic organizations, civil society organizations[7] (CSOs), and local

[5] Yamagata (2016) discussed this aspect in more detail. UNEP and OHCHR (2015) attempted to soothe concerns that the universality principle of the SDGs "undermines the focus on the poorest.".

[6] For example, see the voluntary national reviews on the implementation of the SDGs made by Japan and the United Kingdom (Government of Japan, 2017, p. 9; UK, 2019, Annex III).

[7] In this chapter, the word CSO is used as an interchangeable term for NGO.

governments are considered to be new contributors of Japan's "development cooperation," which supplements the central government of Japan. The resolution of the SDGs also explicitly stresses the role of the private sector, CSOs, and philanthropic organizations as important actors to be involved in a global partnership to achieve the SDGs (UNGA, 2015).

A conclusion from the analyses of the above two sections is that when the SDGs came into effect in September 2015, Japan had already anticipated the three most important traits of the SDGs with the Development Cooperation Charter, which was adopted in February 2015. Japan created the new concept of international cooperation as "development cooperation," which includes peace and security issues as a field of cooperation, middle- high-income countries with vulnerabilities as a beneficiary, and the private sector and civil society as contributors. The SDGs also augment the MDGs in the same three dimensions, i.e., field, beneficiary, and contributor. In this sense the SDGs may be highly agreeable to the Government of Japan, which stepped forward in the same direction a half-year before with the Development Cooperation Charter.

3.2 Institutional Framework to Advance the SDGs

3.2.1 SDGs Promotion Headquarters

The Government of Japan facilitated the implementation of the SDGs with the initiative to create the SDGs Promotion Headquarters, which is headed by the prime minister and composed of all ministers (Government of Japan, 2017, p. 7). Prime Minister Shinzo Abe established the headquarters in May 2016.

Having the prime minister as chief of the headquarters may make it seem that Japan places high priority on the achievement of the SDGs. However, appointing the prime minister as chief of a headquarters or chair of committees in the Cabinet Secretariat is not particularly unusual in Japan at present. Abe's administration was known to concentrate power in the Cabinet Secretariat. The Cabinet Secretariat tends to take initiative for any functions of the government. This control is materialized through appointments of key bureaucrats of ministries in the central government (Burrett, 2017, pp. 415–417). The Cabinet Bureau of Personnel Affairs, which was created in 2014 inside the Cabinet Bureau, designates a hundred of top officials of ministries.

Such a high centralization of power with the prime minister and the Cabinet Secretariat is reflected by the width of coverage of committees,

Table 3 Number of committees, headquarters, and offices that the Cabinet Secretariat organizes

Degree of involvement of the prime minister	*Number*	*Percentage*
Total	261	100.0
Of which the prime minister heads or chairs	51	19.5
Over which the prime minister presides	27	10.3

Source: the website of the Cabinet Secretariat at https://www.cas.go.jp/jp/seisaku/index.html
Note: The above site is written in Japanese. There is no counterpart English page available. The committees over which the prime minister presides are headed or chaired by experts outside the government even though the prime minister is supposed to be present

headquarters, and offices placed inside the Cabinet Secretariat. Table 3 shows that the number of such committees, headquarters, and offices amounts to 261. Out of 261, 51 of those are headed or chaired by the prime minister, while 27 of those are headed by someone else but presided over by the prime minister. Consequently, the prime minister has to be present at the meetings of 78 committees and headquarters. The topics and responsibilities of 261 committees, headquarters, and offices are so diverse as to include abduction issues with the Democratic People's Republic of Korea, administrative reform, aging, agriculture, artificial intelligence technology, decentralization of the public sector, disability, disaster prevention, foreign workers, free trade agreements, global warming, industrial competitiveness, infectious diseases, infrastructure, intellectual property rights, marine plastic trash, national security, nuclear hazard, postal service privatization, rural development, social security, space development, the Tokyo Olympics and Paralympics, tourism, women's empowerment, and so on.

The SDGs Promotion Headquarters is one of 51 committees and headquarters that the prime minister directly heads. Since the prime minister presides over so many meetings of these committees and headquarters, the frequency of meetings of the SDGs Promotion Headquarters has been twice a year only, in June and December since the headquarters was founded in spring 2016. As of August 2020 seven meetings had been held. The June 2020 meeting was canceled because of the new coronavirus outbreak in Japan. Thus, the prime minister's apparently firm commitment to the SDGs as chief of the headquarters makes the frequency of its meetings predictably seasonal and leads Japan's efforts toward the SDGs to appear less dynamic and responsive.

Table 4 Eight priority areas of Japan's SDGs implementation guiding principles (original version released in 2016)

1	Empowerment of all people	5	Energy conservation, renewable energy, climate change countermeasures, and sound material-cycle society
2	Achievement of good health and longevity	6	Conservation of the environment, including biodiversity, forests, and the oceans
3	Creating growth markets, revitalization of rural areas, and promoting science technology and innovation	7	Achieving peaceful, safe, and secure societies
4	Sustainable and resilient land use, promoting quality infrastructure	8	Strengthening the means and frameworks for the implementation of the SDGs

Source: the website of the SDGs Promotion Headquarters; available at https://www.kantei.go.jp/%20jp/singi/sdgs/%dai2/siryou1e.pdf

3.2.2 Duplicity of the SDGs Implementation Guiding Principles

Japan submitted a voluntary national review at the high-level political forum held in July 2017 (Government of Japan, 2017). In order to prepare for the report, the SDGs Promotion Headquarters approved the SDGs Implementation Guiding Principles[8] at the second meeting of the headquarters in December 2016. A detailed action plan for the guiding principles was provided at the same time under the title "Specific Measures to Achieve the Sustainable Development Goals."[9]

Eight priority areas for Japan's SDGs Implementation Guiding Principles are exhibited in Table 4. Roughly speaking, the eight areas are in accord with the 17 SDGs. Priority area 1, the empowerment of all people, covers poverty, education, gender, labor, and inequality (SDGs 1, 4, 5, 8, and 10), while priority area 2 deals with health and nutrition (SDGs 2 and 3). Priority areas 3 and 4 collectively correspond to SDGs 8, 9, and 11,

[8] This original version of the SDGs Implementation Guiding Principles is available from https://www.kantei.go.jp/jp/singi/sdgs/dai2/siryou1e.pdf. The guiding principles were revised in December 2019. This revised edition is available from https://www.kantei.go.jp/jp/singi/%20sdgs/pdf/jisshi_shishin_%20r011220e.pdf.

[9] The "specific measures" are available at https://www.kantei.go.jp/jp/singi/sdgs/dai2/siryou2e.pdf.

while priority areas 5 and 6 are associated with SDGs 6, 7, and 12–15. Finally, priority areas 7 and 8 correspond to SDGs 16 and 17, respectively.

At a glance through Table 4, the beneficiaries of Japan's SDG implementation do not seem to be specified. Some readers may interpret the beneficiaries as people all over the world. However, the original Japanese version of the guiding principles adopted in 2016, and the "specific measures" contain two words implying that the main beneficiaries of Japan's SDGs implementation are the Japanese.

The first word corresponds to "all people" in priority area 1 of Japan's SDGs implementation (Table 4). Under the priority area "the empowerment of all people," various policy measures are listed in the document under "specific measures." The first policy measure included in priority area 1 is "Japan's plan for the dynamic engagement of all citizens,"[10] which was designed to empower all members of Japanese, for instance the aged, youth, children, women, and the disabled. The symbolic terminology incorporated in the original Japanese title of this policy is that the phrase "one hundred million" is used to indicate "all citizens." The direct translation of the original title of "Japan's plan for the dynamic engagement of all citizens" is "Nippon one hundred million people all dynamic engagement plan."

The current population in Japan is estimated at around 120 million. However, one hundred million has been the implied Japanese population for a long time. In the first half of the 1940s, while World War II was in progress, Japan colonized Taiwan, the Korean Peninsula, and the northeast part of China, which was called Manchuria. The total population of these colonies and Japan was around one hundred million at that time. Therefore, "one hundred million people" implied all Japanese and was used in slogans appealing to the solidarity of people in both Japan and its colonizing areas for war support. After two decades passed since Japan's defeat in World War II, Japan's population reached around one hundred million in the mid-1960s. Since then, the phrase "one hundred million" has been used to imply all Japanese. The current world population is approximately seven billion, but one hundred million is the scale at

[10] The details of this policy are available at http://www.kantei.go.jp/jp/singi/ichiokusoukatsuy%20aku/pdf/plan2.pdf.

which beneficiaries are mentioned in documents relating to Japan's SDGs implementation.[11]

The second word that limits beneficiaries to only the Japanese is "land" in priority area 4. In the guiding principles in Japanese, a word meaning "national land" is used, while the adjective "national" is omitted in the English version of the guiding principles. The original phrase "resilient land use" in the guiding principles implies "building national resilience" according to the Cabinet Secretariat.[12] This concept is based on the Basic Act for National Resilience Contributing to Preventing and Mitigating Disasters for Developing Resilience in the Lives of the Citizenry.

These two examples reveal that there are duplicities between the original Japanese version and the translated English version of the guiding principles. While the Japanese version is aimed at convincing the Japanese that the main beneficiaries of Japan's SDGs implementation are the Japanese, its translation into English obscures the own-country tone of the guiding principles.

Finally, there is another indication that Japanese beneficiaries are greatly prioritized over global beneficiaries even though the SDGs are supposed to be global goals. That indication lies in the document entitled "Specific Measures to Achieve the Sustainable Development Goals." This document consists of tables summarizing (1) Japan's policy measures, (2) corresponding SDG targets, (3) performance indicators, and (4) the responsible Japanese ministry or agency according to Japan's priority areas to promote the SDGs. The policy measures are further divided into domestic measures and overseas measures. Roughly speaking, the domestic measures are planned for the sake of the Japanese, while the overseas measures are designed for all people on this planet. The document on "specific measures" is 46 pages long. Throughout the document domestic measures are always placed ahead of overseas measures. Moreover, out of the 46 pages, about 18 pages are allocated to the overseas

[11] The SDGs Implementation Guiding Principles were revised in 2019. In the revised guiding principles priority area 1 was changed from the empowerment of all people to "the realisation of gender equality and a society where every person can play an active role." Though the phrase "one hundred million people" was mentioned in the Japanese version of the original guiding principles, it no longer appears in the Japanese version of the revised guiding principles. See https://www.kantei.go.jp/jp/singi/sdgs/pdf/%20jisshi_shishin_r011220e.pdf for the revised guiding principles.

[12] See the website of the Cabinet Secretariat for "building national resilience": https://www.cas.go.jp/jp/seisaku/kokudo_kyoujinka/index_en.html.

measures. This implies that the ratio between domestic and overseas measures is around 60 to 40. Since the policy measures associated with priority area 7 are all overseas measures, the domestic to overseas ratio becomes far higher for priority areas 1 to 7. The bottom line is that Japan's policy measures to promote the SDGs are domestic-measure-dominant, which emphasizes that the main beneficiaries of Japan's efforts to achieve the SDGs are the Japanese.

3.2.3 Japan SDGs Award

One of the salient efforts that the Government of Japan has made toward the SDGs is the creation of the Japan SDGs Award. This is highlighted on the home page of the SDGs Promotion Headquarters and Japan's voluntary national review (Government of Japan, 2017, p. 5). The Secretary-General of the United Nations, António Guterres, sent congratulatory messages for the first and second presentation ceremonies of this award.

This award was created to "encourage the private sector to work on the SDGs" in 2017 (Government of Japan, 2017). In the first year, 12 Japanese private firms, local governments, and nonprofit organizations (NPOs) were granted the award. Likewise, 15 companies, NPOs, and local governments earned the award in 2018, while 11 organizations were awarded in 2019. The fourth award for 2020 was accepting applications as of August 2020.

This award symbolizes the attitude of the Government of Japan toward mobilizing foreign pressure (gaiatsu in Japanese) to achieve domestic goals.[13] The SDGs can be interpreted as foreign pressure. This foreign pressure is transformed into energy to promote the local development of the Japanese economy through the following mechanisms.

Firstly, applicants for the Japan SDGs Award are limited to firms or organizations located inside Japan. This limitation is outlined in the application manual, which is available in Japanese only.[14] The limitation is

[13] Making use of foreign pressure (*gaiatsu*) to justify a certain domestic policy is a typical attitude of the Government of Japan. Lancaster (2010) applied this *gaiatsu* hypothesis to Japan's ODA policy.

[14] The manual is found at http://www.kantei.go.jp/jp/singi/sdgs/japan_sdgs_award_dai4/02jisshi.pdf.

not spelled out in any documents explaining the award in English.[15] Above all, no announcement regarding applications written in English is produced. The Japan SDGs Award is a prize for the Government of Japan to praise Japanese organizations and firms.

Secondly, domestic contributions are more highlighted than overseas contributions. The Japan SDGs Award is composed of three categories. The foremost award is called the "Chief's Award," and it is granted by chief of the headquarters, namely the prime minister of Japan. There are two deputy chiefs of the headquarters, the Chief Cabinet Secretary and the Foreign Minister. The secondary awards are presented by the two deputy chiefs. Broadly speaking, the Chief Cabinet Secretary grants the award to organizations engaged in local development or sustainability in Japan, while the Foreign Minister does so for those contributing to international development in developing countries. The third award falls under the title "SDGs Partnership Award." All three times the awards have been granted since their establishment, awardees of the first prize presented by the prime minister were organizations or local governments that contributed to local development and sustainability. So far, none featuring international development have been selected for the Chief's Award.

It is a pity that the Japan SDGs Award is not something Japan uses to encourage all people in the world to contribute to the SDGs. Not only the organizers but also the awardees of this prize are destined to be Japanese.[16] Because of the universality principle of the SDGs, this inward-oriented nature of the Japan SDGs Award does not seem to be accommodated by the SDGs.

4 Awakened Private Sector to the SDGs

In Japan the SDGs are promoted by the government, the private sector, and civil society. The predecessors of the SDGs, namely the MDGs, were advocated by civil society and the Government of Japan in the context of international cooperation. Since the beneficiaries of the MDGs were limited to those living in developing countries, most of the Japanese

[15] Japan's voluntary national review submitted in 2017 (Government of Japan) and the website of the Ministry of Foreign Affairs on this award (https://www.mofa.go.jp/policy/oda/sdgs/award/index.html) also do not list this eligibility criterion.

[16] This is not in terms of nationality, but the location of organizations and firms.

private sector did not respond to the MDGs. However, under the SDGs, local sustainable development in Japan could be commended by the international community. The private sector took this as an opportunity to encourage the exercise of corporate social responsibility (CSR).

A key actor that leads the private sector toward the SDGs is Keidanren (Japan Business Federation), which is an umbrella association of "1,444 representative companies of Japan, 109 nationwide industrial associations and the regional economic organizations for all 47 prefectures (as of April 1, 2020)" according to its home page.[17] Indeed, Keidanren is an important opinion leader of the private sector of Japan.

Keidanren operates under the Charter of Corporate Behavior, which binds member firms and associations to its observation. In 2017 Keidanren revised this charter and incorporated the SDGs into it. The revised charter has the subtitle "For the Realization of a Sustainable Society." It is impressive that though Keidanren did not respond to the United Nations Global Compact[18] when then Secretary-General of the United Nations Kofi Annan visited Keidanren in 2004,[19] it fully committed to the SDGs in 2017.

Keidanren's conversion to the SDGs is due to the compatibility of the SDGs with innovation and local development in Japan, which originated from the universality principle. The preamble of the revised Charter of Corporate Behavior is entitled "Delivering on the SDGs through the Realization of Society 5.0." Society 5.0 is defined as a future society adopting IoT, AI, robots, and other innovative technologies.[20] Keidanren seems to find the SDGs to be a close match for Society 5.0. The pursuit of Society 5.0 in Japan can be interpreted as a contribution to the achievement of the SDGs. Because the number of SDGs is 17, which is arguably too many to realize all at once, no one is obliged to meet all 17 goals

[17] The cited page is the following: https://www.keidanren.or.jp/en/profile/pro001.html.

[18] The United Nations Global Compact is an initiative to let global firms agree on a code of conduct for human rights, labor, environment, and anticorruption. Kofi Annan established it in 2000.

[19] See Keidanren's weekly magazine, *Keiei Times*, no. 2712, 4 March 2004 (in Japanese). Available at http://www.keidanren.or.jp/japanese/journal/times/2004/0304/01.html.

[20] Keidanren (2017) defines Society 5.0 as "the fifth and newest society in the history of human social development, following on from the hunter–gatherer society, agrarian society, industrial society, and information society.".

at the same time. If one contributes to a part of them, that partial achievement deserves applause.[21]

Keidanren is now a forerunner of the SDGs on behalf of the Japanese private sector. It maintains a website entitled KeidanrenSDGs[22] and introduces good practices to its member firms and the general public. Major Japanese mass media follow Keidanren. Many Japanese newspapers and TV broadcasting firms became members of the SDG Media Compact[23] of the United Nations, which supports the SDGs as mass media. The Nikkei holds seminars and events such as the "Nikkei SDGs Festival" and the "Nikkei SDGs Forum" to encourage commitments to the SDGs. It introduces good practices for firms through its website.[24] *The Asahi Shimbun* also actively reports on the SDGs. It maintains a website for "2030 Transform through SDGs,"[25] which summarizes SDG-related news and activities. A group company for satellite TV of *The Asahi Shimbun*, BS Asahi, broadcasts a weekly program on good practices for the SDGs.[26] Two other major newspaper publishers, *The Mainichi* and *The Yomiuri Shimbun*, also have websites featuring the SDGs.[27]

Since the Government of Japan submitted its voluntary national review in July 2017, Japan's commitments to the SDGs have been strengthened. Keidanren revised its Charter of Corporate Behavior in November 2017. Then, mass media followed to advertise the SDGs. This is the process of the SDGs becoming enshrined by the private sector, NPOs, and local administrations in Japan; however, the SDGs were overshadowed by global warming in 2019 and COVID-19 in 2020 in terms of momentum within the United Nations.

[21] This is a view given by Keidanren (2017b, p. 142). Details of this view are given in Section V.

[22] See https://www.keidanrensdgs-world.com/.

[23] See https://www.un.org/sustainabledevelopment/sdg-media-compact-about/.

[24] See https://r.nikkei.com/stories/topic_DF_TH_20030200.

[25] See http://www.asahi.com/ajw/special/SDGs/.

[26] See https://www.bs-asahi.co.jp/passthebaton/.

[27] See https://mainichi.jp/ch190435490i/SDGs for *The Mainichi* and https://kyoiku.yomiuri.co.jp/sdgs/ for *The Yomiuri Shimbun*.

5 Concluding Remarks

The SDGs are popular in Japan even with the worldwide COVID-19 outbreak from the end of 2019. For the private sector and local governments, the SDGs are useful as a window for them to achieve CSR, sustainability, globalization, and local development in Japan.

While the SDGs are becoming well known by the Japanese, international development could be left behind. For many Japanese the SDGs are divided into a range of areas that is too wide. Indeed, Keidanren recommends that its member firms place high priority on some favorable ones out of the 17 SDGs, according to the direction of its business activities, because it is difficult for a firm to be engaged in all 17 goals.[28] This implies that it is likely for a firm to select a few easily achievable goals, which are unlikely to be those targeting developing countries.

The SDGs and international development are decoupled. This tendency is accentuated in Japan where the SDGs are advocated by firms and organizations whose activities are mostly undertaken inside Japan. There seems to be no concern that the predecessors of the SDGs were the MDGs, the beneficiaries of which were people in developing countries. Thus, in Japan international development is being isolated from the SDGs. The achievement of (some of) the SDGs need not lead to poverty reduction in developing countries.

References

Arase, D. (1995). *Buying power: The political economy of Japan's foreign aid*. Lynne Rienner Publishers.

Burrett, T. (2017). Abe road: Comparing Japanese prime minister Shinzo Abe's leadership of his first and second governments. *Parliamentary Affairs, 70*(2), 400–429. https://doi.org/10.1093/pa/gsw015

Cabinet. (2015, 10 Febrauary). *Development cooperation charter: For peace, prosperity and a better future for everyone*. Ministry of Foreign Affairs, Government of Japan. https://www.mofa.go.jp/files/000067701.pdf. Accessed 5 April 2021.

[28] This recommendation is given in a column entitled "Ways to incorporate the SDGs into corporate management" in Keidanren's manual, *Implementation Guidance on the Charter of Corporate Behavior* (Keidanren, 2017b, p. 142). *Guidance* is published only in Japanese. However, its table of contents in English is posted at https://www.keidanren.or.jp/en/policy/csr/tebiki7.html.

Commission on Human Security. (2003). *Human security now*. Commission on Human Security. https://reliefweb.int/sites/reliefweb.int/files/resources/91BAEEDBA50C6907C1256D19006A9353-chs-security-may03.pdf. Accessed 5 April 2021.

Ensign, M. M. (1992). *Doing good or doing well?: Japan's foreign aid program*. Columbia University Press.

Government of Japan. (2017). *Japan's voluntary national review. report on the implementation of the sustainable development goals*. SDGs Promotion Headquarters. https://www.mofa.go.jp/files/000287390.pdf. Accessed 5 April 2021.

Japan International Cooperation Agency (JICA) (Ed.). (1998). *The OECD/DAC's new development strategy: Report of the issue-wise study committee for Japan's official development assistance* . JICA. https://www.jica.go.jp/jica-ri/IFIC_and_JBICI-Studies/english/publications/reports/study/topical/oecd_dac/pdf/dac1.pdf. Accessed 5 April 2021.

Kamidohzono, S. G., Gómez, O. A., & Mine, Y. (2016). Embracing human security: New directions of Japan's ODA for the 21st century. In H. Kato, J. Page, & Y. Shimomura (Eds.), *Japan's development assistance: Foreign aid and the Post-2015 agenda* (pp. 205–221). Palgrave.

Keidanren. (2017a). *Charter of corporate behavior* (provisional translation). Keidanren. https://www.keidanren.or.jp/en/policy/csr/charter.html. Accessed 5 April 2021.

Keidanren. (2017b). *Kigyo kodo kensho: jikko no tebiki* (in Japanese: Implementation guidance on the Charter of Corporate Behavior), 7th ed. Keidanren. http://www.keidanren.or.jp/policy/cgcb/tebiki7.pdf. Accessed 5 April 2021.

Lancaster, C. (2010). Japan's ODA: naiatsu and gaiatsu. Domestic sources and transnational influences. In D. Leheny & K. Warren (Eds.) *Japanese aid and the construction of global development: Inescapable solutions* (pp. 29–53). Routledge.

Leheny, D., & Warren, K. (2010). Introduction. In D. Leheny & K. Warren (Eds.), *Japanese aid and the construction of global development: Inescapable solutions* (pp. 1–26). Routledge.

Ministry of Foreign Affairs (MOFA), Government of Japan. (1997). *Japan's ODA annual report (summary) 1997*. Government of Japan. https://www.mofa.go.jp/policy/oda/summary/1997/index.html. Accessed 5 April 2021.

Ministry of Foreign Affairs (MOFA), Government of Japan. (2003). Japan's Official Development Assistance Charter. *MOFA*. https://www.mofa.go.jp/policy/oda/reform/revision0308.pdf. Accessed 5 April 2021.

Organisation for Economic Co-operation and Development, Development Assistance Committee (OECD/DAC). (1996a). *Shaping the 21st century: The contribution of development co-operation*. OECD. https://www.oecd.org/dac/2508761.pdf. Accessed 5 April 2021.

Organisation for Economic Co-operation and Development, Development Assistance Committee (OECD/DAC). (1996b). *Development co-operation review series, Japan*, No. 13. OECD.

Organisation for Economic Co-operation and Development, Development Assistance Committee (OECD/DAC). (1999). *Development co-operation review series, Japan, 1999*, No. 34. OECD.

Orr, R. M., Jr. (1990). *The emergence of Japan's foreign aid power*. Columbia University Press.

Open Working Group on Sustainable Development Goals (OWG-SDGs). (2014). *Encyclopedia groupinica: A compilation of goals and targets suggestions from OWG-10*. United Nations. https://sustainabledevelopment.un.org/content/documents/3698EncyclopediaGroupinica.pdf. Accessed 5 April 2021.

United Kingdom of Great Britain and Northern Ireland (UK). (2019). *Voluntary national review of progress towards the sustainable development goals*. HM Government. https://www.gov.uk/government/publications/uks-voluntary-national-review-of-the-sustainable-development-goals. Accessed 5 April 2021.

United Nations Development Programme (UNDP). (2012, 13 March). East Asian regional consultation on the Human Development Reports and measurement of progress. Paper presented at the meeting *East Asian Regional Consultation on Human Development Report and Measurement of Human Development Progress*. JICA Research Institute.

United Nations Environment Programme (UNEP) and Office of the United Nations High Commissioner for Human Rights (OHCHR). (2015). *Universality in the post 2015 sustainable development agenda*. UNEP Post-2015 Note #9. UNEP. https://www.ohchr.org/Documents/Issues/MDGs/Post2015/OHCHR_UNEP.pdf. Accessed 5 April 2021.

United Nations General Assembly (UNGA). (2013). *Open working group of the general assembly on sustainable development goals. Resolution at the 67th session of UNGA*. A/67/L.48/Rev.1. United Nations. https://digitallibrary.un.org/record/%20742043/files/A67L.48_%20Rev.1-EN.pdf. Accessed 5 April 2021.

United Nations General Assembly (UNGA). (2014). *Report of the open working group of the general assembly on sustainable development goals*. Resolution at the 68th session of UNGA. A/68/970. United Nations. https://www.un.org/ga/search/viewdoc.asp?symbol=A/68/970&Lang=E. Accessed 5 April 2021.

United Nations General Assembly (UNGA). (2015). *Transforming our world: The 2030 agenda for sustainable development*. Resolution at the 70th session of UNGA. A/70/L.1. United Nations. https://www.un.org/ga/search/view_doc.asp?symbol=A/RES/70/1&Lang=E. Accessed 5 April 2021.

World Health Organization (WHO). (2013). *The world health report 2013: Research for universal health coverage*. WHO. https://www.afro.who.int/sites/default/files/2017-06/9789240690%20837_eng.pdf. Accessed 5 April 2021.

Yamagata, T. (2016). Sustainable development goals and Japan: Sustainability overshadows poverty reduction. *Asia-Pacific Development Journal, 23*(2), 1–17. https://doi.org/10.18356/5d804069-en

CHAPTER 12

A Way Forward to Achieve the SDGs in Korea: Reformulating an Enabling Environment for Sustainable Development

Taekyoon Kim and Bo Kyung Kim

1 Introduction

Transformation from the Millennium Development Goals (MDGs) to the Sustainable Development Goals (SDGs) was a reflexive change aiming for the "*re*-structuration and *re*-conceptualization" of existing development goals rather than "*de*-structuration and *de*-conceptualization" (emphasis in original) (Beck et al., 2003, 3). The kernel of the SDGs is their expanded multidimensional approach, which supplemented the former MDGs that were narrowly centered on social development. The comprehensive and universal features of the new agenda require participation

T. Kim (✉) · B. K. Kim
Seoul National University, Seoul, South Korea
e-mail: oxonian07@snu.ac.kr

B. K. Kim
e-mail: bk.kim@snu.ac.kr

H.-j. Kwon et al. (eds.), *International Development Cooperation of Japan and South Korea*,
https://doi.org/10.1007/978-981-16-4601-0_12

from all levels of actors, since it interprets development as a holistic concept that encompasses not only social but also economic and environmental aspects. Yet national implementation of the goals is still regarded as challenging for both the Global North and South, since the process is by and large a matter of state autonomy depending on their respective governance capacity. Here, governance and partnership, as specified in goals 16 and 17, are key principles that crosscut the entire agenda. The underlying assumption is that improving the governance mechanism is both a means of implementation and an end to ultimately achieve sustainable development in the long run (Meadowcroft, 2010; Monkelbaan, 2019).

Governance can be a tricky term that has been diversely defined in multi-level forms spanning from local to national and global. Moreover, the polysemy of the concept has generated the idea that governance goes nowhere beyond an extended term that lacks common denominators; thereby, it can explain everything and yet explain nothing (Jessop, 2003; Offe, 2009; Torfing et al., 2012). In particular, it is a challenging task to blend global norms into domestic policies as there cannot be a unified single method transferrable from one country to another. At times, implementation itself can become stalled when confronted with resistance. Hence, prior thought must be given to finding a middle ground between hierarchical governance and a bottom-up network approach with third sector participation. The importance of a bottom-up model of policy change goes hand in hand with the inclusive features of SDGs that actively invite actors of the third sector into the development field. These include nongovernmental organizations (NGOs), businesses, private foundations, civil society organizations (CSOs), and many other entities that play roles and functions related to service delivery and advocacy with a rights-based approach, fund provision for development cooperation, and oversight as watchdogs (Lewis, 2001, 2014). Represented by NGOs and CSOs, the existence of the third sector is essential to attaining a whole-of-society network approach in implementing global agendas in the national context.

Recognized in earnest since the 2008 High-Level Forum on Development Effectiveness in Accra, third sector participants were officially proclaimed to be essential agents of development cooperation. The existence and importance of the third sector have been underscored as they were to be regarded as "development actors in their own right, with their own priorities, programmes and partnership arrangements" (OECD,

2009, 13). As implementation of the SDGs is heavily dependent on how well countries use key tools for the nationalization of the goals, the role of the third sector has gained more credibility, with its accessibility to grassroots knowledge and capacity to take quick action against government wrongdoing (Hedman & McDonnell, 2011; Kim, 2017; OECD, 2012b). The 2011 Fourth High-Level Forum on Aid Effectiveness held in Busan proved that enlarging participation from the third sector is a core element to enhance policy coherence and development effectiveness. Here, the Istanbul Principles for CSO Development Effectiveness and the International Framework for CSO Development were officially endorsed (OECD, 2011, 6).

Under such a changing global atmosphere that began to emphasize multi-stakeholder cooperation at all levels, promoting and encouraging partnerships among public, public–private, and the third sector became important preconditions for implementing the goals, as also indicated under SDG 17.17.[1] In this context, creating a symbiotic relationship between the government and the third sector has been regarded as essential to pioneering ways toward a whole-of-society approach for achieving sustainable development (Seery & Seghers, 2019). Nevertheless, incorporating the third sector can be difficult in countries where civil society agents lack the capacity to fulfill advocacy functions or to collaborate with international networks and organizations. Guidelines on how to maximize enablers and minimize disablers naturally differ from country to country depending on their pregiven conditions, such as institutional capacity, organized grassroots movements, and the existence of advocacy coalitions. South Korea is an unmatched example in terms of its development despite the relatively short decade-long experience of becoming an OECD DAC member since 2010. The course of evolution of the third sector and its partnership-building with the government along the stream of policy change have shown great dynamism. As a counterpart in partnership with the third sector, the South Korean government has also made great efforts to build institutional platforms and frameworks for the improvement of policy coherence through civil participation.

South Korea also showcases how a whole-of-society approach can be applied to achieving policy coherence for sustainable development

[1] SDG target 17.17 is as follows: "Encourage and promote effective public, public–private and civil society partnerships, building on the experience and resourcing strategies of partnerships."

(PCSD).[2] Throughout its OECD DAC peer reviews, the issue of institutional fragmentation has been continuously pointed out as a key vulnerability that hinders the country from fully adopting the framework (OECD, 2012a, 2018). Looking into the PCSD framework, its core idea has developed within the OECD over two decades, creating greater potential for application to the post-2015 development agenda. As a global mechanism to form an enabling environment for achieving the 2030 Agenda for Sustainable Development, the original policy coherence for development (PCD) has transformed into a more inclusive and broader approach since 2014. Whereas the PCD focused on how donor policies affect recipients on a unilateral basis, the PCSD has evolved into an implementation mechanism whereby the government and the third sector create synergies and complement each other to achieve shared goals.[3] The importance of involving the third sector as key stakeholders has been stressed with the identification of five complementary levels of coherence for implementation of the SDGs (OECD, 2016a, 88).[4]

The dominant approach when it comes to explaining modes of achieving such multi-layered levels of coherence is that a strong bottom-up approach is relatively ideal to establish more inclusive partnerships (Sabatier, 1986; Woo et al., 2019, 275). In this regard, comprehensive implementation of the SDGs requires new enabling environments that allow transformation in governance mechanisms for social integration (Woo et al., 2019). Dealing with such issues, this chapter first outlines hierarchical and network approaches to governance and highlights the need for a middle ground with an advocacy coalition framework. In line

[2] The OECD toolkit for self-assessment on institutional mechanisms and practices on policy coherence for sustainable development (PCSD) includes the following checklist items: (1) policy commitment, (2) long-term strategic review, (3) policy integration, (4) policy coordination, (5) local and regional involvement, (6) stakeholder engagement, (7) policy and financial impacts, and (8) monitoring, reporting, and evaluation.

[3] SDG target 17.14 is as follows: "Enhance policy coherence for sustainable development." Its global indicator 17.14.1 takes into account the number of countries with mechanisms in place to enhance policy coherence for sustainable development.

[4] The five levels of coherence include the following: (1) between the SDGs and national policies (vertical coherence); (2) between the 2030 Agenda for Sustainable Development and other international agendas; (3) between economic, social, and environmental policies; (4) between diverse sources of finance; and (5) between actions of multiple actors (governments, international organizations, civil society, and the private sector) (OECD, 2016a, 88–89).

with PCSD, related concepts are reviewed to navigate conditions for an enabling environment. Then, the chapter goes on to examine the case of South Korea, where government policies on SDGs implementation have evolved with the growth of the third sector. How the goals were nationalized into the Korean Sustainable Development Goals (K-SDGs) and the development of a partnership between the government and the third sector are surveyed. Lastly, by summarizing the policy direction of South Korea toward embracing the whole-of-society approach, the chapter concludes with discussions on challenges ahead.

2 Modes of Governance and Advocacy Coalition

Among several approaches to governance seeking social coordination, two can be considered applicable when implementing the SDGs. One is strong top-down leadership based on a monocentric hierarchical system; another is bottom-up inclusive participation based on policy community and a network (Meuleman, 2018). The hierarchy approach is legally binding and links to terms such as authority, bureaucracy, and prearranged planning by the government through top-down policy measures (Jessop, 1997). The approach argues that in the case of conflicts of interest among groups within a society, social coordination can be forced by explicit or implicit political authority capable of taking forceful actions if deemed necessary (Chisholm, 1989; Lindblom, 1977; Rhodes, 2007; Scharpf, 1997). Under the hierarchical approach to governance, a holistic plan is designed to solve social problems in an intended direction, and central leadership at the national level is regarded as one of the key success factors for development. The centralized leadership understands that policy and governance are terms with kindred meanings, which are realized through SDG 16 that emphasizes the institutional aspect (Woo et al., 2019).

On the other hand, the network approach goes along with terms such as trust and community, with voluntarily organized communities being self-reflexive and membership being socially binding (Jessop, 1997). The approach insists that there should be a strong bottom-up partnership with a mutually agreed coercion mechanism that is autonomously enforced. Although the definition of mutual agreement is not as strict as it is according to Lijphart's (1999) majoritarian consensus dichotomy, biased options naturally create restraints under the network approach (Hardin, 1968). Similar to the hierarchical system, an individual's freedom can be violated to a significant degree once agreed by members of the network.

The difference lies in that coercion does not stem from authority, but instead from common norms, values, and mutual consent. Although it does not rely on legal force, political authority rules over networks, and the network functions as a mode of social coordination in place of such political authority based on voluntary cooperation and autonomous compulsion (Beetham, 1996). Ostrom (1990) finds that problem-solving through voluntary cooperation can be more efficient than government intervention. Her argument supports the idea that when proper conditions are met, the public can successfully solve collective problems revolving around the common good, even without arbitration by the authority.

In this regard, social coordination to implement the SDGs can be analyzed with the network approach, which involves voluntary and horizontal cooperation based on self-regulation enforced by mutual consent. This specific mode of coordination is advantageous in overcoming the problem of limited rationality and opportunistic behavior. As an ideal form of social coordination, a network approach to governance has features of trust, mutual understanding, reciprocity, informality, cooperation, mutual adjustment, shared ethical or moral commitment, a sense of common purpose, and trustworthy communication. In particular, "trust" can be regarded as the core element of network governance in that shared values and norms are an adhesive that connects the complex relationships among networks (Rhodes, 1997, 2000; Sørensen & Torfing, 2005; Torfing, et al., 2012). However, in order for such a network approach to be realized to its fullest extent, a significantly high level of trust is required for it to function and allow inclusive coordination among interest groups. Not withstanding that the network approach overly relies on voluntarism, such risk can be canceled out by either economic incentives or the existence of a hierarchical system that can intervene whenever horizontal cooperation does not function (Jessop, 1997; Meuleman, 2018).

The core is the problem-solving aspect of governance. Seeking transferability of coordination mechanisms, the difficulty of implementing the SDGs is that coordination is still based on a top-down approach, while governance through inclusive participation is being increasingly emphasized (Common, 2013; United Nations, 2018, 2019). With such an understanding, the goals of the SDGs can be taken as the "commons" in Ostrom's terms (1990), and the network approach to governance supplemented by a hierarchical system can therefore be understood as a mechanism for social coordination. In this context, seeking long-term policy change from top-down toward bottom-up methods requires a buffer of an intermediary stage at which a middle ground for both approaches can coexist during the transition. As Kuenkel (2017) finds, dynamic balance only comes after transformation, and during times of transition, there is plurality. In other words, as a path for creating an enabling environment for policy coherence, there needs to be a foothold between hierarchical and network approaches.

This transitional phase can be explained with the advocacy coalition framework (ACF) from the policy change literature. The ACF brings together both top-down approaches whereby policy decisions are handed down to local levels and bottom-up approaches whereby local participation greatly impacts the policy-making process on upper tiers (Sabatier, 1988, 158). In theory, policy change occurs with gradual transformations that are accompanied by new strategies. It also requires a robust institutional foundation that enables such coalitions to be represented as policy subsystems (Bennett & Howlett, 1992). Long-term planning is required for policy change unless there is a catalyst that creates incentives toward radical reform. Often, change does not occur at all, with actors who prefer maintenance over transformation and with institutions that tend to be too rigid and slow to evolve (Mahoney, 2000; Pierson, 2000). As it is useful in explaining policy areas involving large numbers of stakeholders and wherein only long-term change can be sought,[5] ACF provides room to explain policy change over long durations and emphasizes policy subsystems of advocacy coalitions as key units (Cerna, 2013; Sabatier & Jenkins-Smith, 1993, 1999). Likewise, to understand how advocacy coalitions are formed as subsystems, it is important to look into how third sector participation is organized.

It is largely agreed that third sector involvement in global governance has a substantial impact in terms of reducing accountability deficits (Scholte, 2011). This is based on the understanding that the third sector greatly contributes to the policy process and ensures transparency of global agencies with its oversight function (Chandler, 2004). Actors in the third sector are key units of development, as they are voluntarily formed with autonomous purposes. As highlighted in the Busan partnership agreement, "CSOs play a vital role in enabling people to claim

[5] Examples include areas such as the environment, education, and energy sectors.

their rights, in promoting rights-based approaches, in shaping development policies and partnerships, and in overseeing their implementation." The Nairobi Outcome Document restated the essential role of civil society as "independent partners" (GPEDC, 2016; OECD, 2011). Identifying incentives that are commonly shared among different stakeholder groups can expand the boundaries of participation levels. The ideal goal is to enter the ultimate stage of an enabling environment with an all-inclusive feature, as shown in Fig. 1.

Third sectors in most countries have reported that governments use civil society elements in their policy dialogue only for the purpose of communicating with partner country governments (Wood & Fällman, 2019). However, in order to seek systemic change, expansion from Area B in Fig. 1 to an enabling environment for all-inclusiveness in Area C requires further recognition of other stakeholder groups. Such stakeholders not only include national governments and CSOs, but also academic institutions, businesses, development partners, local government and authorities, epistemic communities as interlocutors, transnational advocacy networks, and many more that enable a whole-of-society

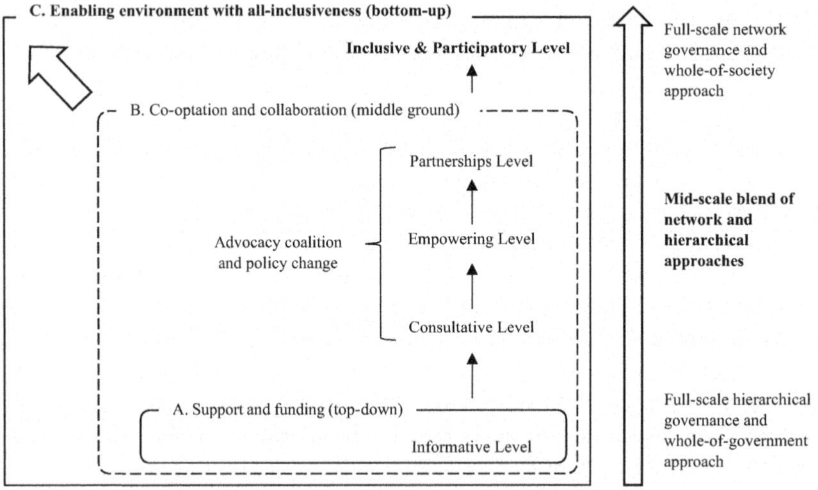

Fig. 1 Expanding boundaries to an enabling environment with all-inclusiveness (*Source* Modified by authors from Kindornay and Zeki [2019, 3])

approach. Epistemic communities have played a significant role in international agenda setting, whereby framing countries' interests are put forward through actual policies designed by a network of expert groups based on expertise and knowledge (Haas, 1989). In addition, environmental NGOs are also diversifying their access to environmental politics through the power and influence of the network as it has developed into a so-called transnational advocacy network whose scope and influence exceeds the scope of respective countries (Keck & Sikkink, 1998). Policy communities (sub-government, issue network, advocacy coalition, etc.) formed around SDGs involve a complex coordination mechanism, as dimensions are not confined to environmental issues. This explains why actors and entities try to form networks when it comes to SDGs instead of acting as agents seeking ways to jump on the bandwagon of so-called policy monopoly (Baumgartner & Jones, 1993).

The international community acknowledges that good governance and democracy without corruption cannot exist without active civil society movements. The majority of the OECD DAC members have been establishing connections with diverse actors in civil society by enacting policies that enable equal participation and partnership. Development partners can promote an enabling environment for the third sector to engage in their advocacy role in a more systemic manner (Arstein, 1969). Methods include considering the third sector as key actors of development, strengthening funding mechanisms that can improve operations, ensuring independence and transparency, and enhancing capacity to respond to local needs (Cázarez-Grageda, 2018; CIVICUS, 2019; P4R, 2018). As such, the active participation of the third sector leads to resilience of public policies that are supplemented with collaborative governance through a whole-of-society approach. However, meaningful coordination across policy sectors cannot be realized without the trade-off of short-term needs for the longer-term ideal. Such difficulties can be overcome if supported by strong institutional foundations that can integrate the policy process, as well as working governance mechanisms that reflect local needs and grassroots voices, specialized knowledge, and hands-on experience (OECD, 2019). The essence of such integration is to ensure inclusiveness and participation from all levels including the third sector, which is why a whole-of-society approach is necessary to create a partnership-enabling environment.

The next section outlines how a whole-of society approach has been realized so far in South Korea to implement the SDGs on the domestic level by examining the following: (1) establishment of the K-SDGs led by the National Commission on Sustainable Development and how the third sector was incorporated along the process of policy change; (2) evolution of government and third sector partnerships in South Korea; and (3) how the shift from partnerships of co-optation to inclusiveness has expanded toward an enabling environment for all-inclusiveness.

3 Whole-of-Society Approach in South Korea for SDGs Implementation

3.1 K-SDGs Led by NCSD and Policy Change[6]

As a central organization for nationalizing the SDGs in South Korea, the National Commission on Sustainable Development (NCSD) was launched in 2000 as a Presidential Advisory Committee, following the declaration of the New Millennium National Environmental Vision. The following year, the Basic Act for Sustainable Development was promulgated and enforced in 2008 to establish a legal mechanism to legitimize domestic policies for sustainable development. Since then, the Basic Act for Sustainable Development has been renamed the Sustainable Development Act, and the NCSD was relocated under the Ministry of Environment in 2010. Adoption of the act was based on Article 50 of the Framework Act on Low Carbon Green Growth, and conflicts of interest among parties were mediated by the NCSD in consultation with regional committees on green growth. As a network organization of central and regional committees, the NCSD consists of a maximum of 50 central committee members, over half of which must be nongovernmental officials. Expert committees supplement research and examination in specialized areas that require policy advice. The key functions of the NCSD are to provide knowledge and information on sustainable development to citizens on behalf of the government and to collect feedback with civil society responses to be reflected in future revisions of the basic plans for sustainable development as implementation strategies.

[6] This section is based on the Basic Act for Sustainable Development, the Sustainable Development Act, the Framework Act on Low Carbon Green Growth (accessed through the Korean Law Information Center website), and official information posted on the website of the National Commission on Sustainable Development (NCSD).

Four implementation plans have been established so far. The purpose of the plans is to revise national strategies every five years for a 20-year planning period in order to implement international agreements on sustainable development in the South Korean context. Matters of concern include the current state of affairs in response to the agenda, changes in conditions, prospects and visions, objectives, strategies, principles, basic policy directions, selecting key targets and indicators, and alignment with international agreements. The First National Sustainable Development Strategy and Implementation Plan (2006–2010) was announced as an integrated management strategy and action plan in the economic, social, and environmental sectors for the first time. A total of 48 areas were addressed in 16 strategies. The Second Basic Plan for Sustainable Development (2011–2015) dealt with national visions and goals in response to changes in the global environment and shifting prospects on sustainable development. Basic policy directions were based on four strategies for 25 implementation tasks, and 77 national sustainable development indicators were devised for analysis. During the earlier stages of the first and second periods, implementation strategies were mostly planned and executed by the government ministries and related offices in a top-down manner.

The Third Basic Plan for Sustainable Development (2016–2020) can be marked as a crucial turning point, as it reflected the UN SDGs according to domestic conditions. Its focus was on addressing vulnerable areas and threat factors based on a national sustainability assessment, with reinforcement of integration among the economic, social, and environmental sectors for advancement from the Second Basic Plan. The Third Basic Plan sets outs 50 implementation tasks in 14 strategies aimed at achieving four objectives: a healthy environment, a safe and integrated society, an inclusive and innovative economy, and a globally responsible country. A total of 122 targets were extracted from the 17 areas with development goals, and 214 indicators were used to ensure policy coherence. Of the 214 indicators, 122 were newly designed to reflect the reality that fits the country context, which clearly shows voluntary efforts to seek a balance between global standards and nation-specific conditions. During this period, the South Korean government established the K-SDGs (Korean Sustainable Development Goals) after "reinforcing sustainable development" was put forth as part of the national agenda in 2018. Unlike the existing top-down method, the establishment of the K-SDGs was promoted with a bottom-up method, with participation not only from the relevant government ministries and offices, but also from private working groups, private stakeholders, and the public.

The most recently adopted Fourth Basic Plan for Sustainable Development (2021–2040) amended the former Third Basic Plan with major changes concentrated on re-listing focus sectors into 119 targets and 236 indicators. Updating the policy objectives and associated indicators that are most relevant to the South Korean context was the core change. Whereas indicators for the third plan were mostly centered around the environmental field, the fourth plan included indicators that could be applied to the social and economic sectors. This created a balanced system for pursuing sustainable development on the national level. While revising the targets and indicators listed as K-SDGs during the third and fourth round, an array of networks was organized for actual goal setting, including the pan-ministerial K-SDGs consultative groups, a private–government–academic joint working group for each goal,[7] and the K-MGoS (Korean Major Groups and other Stakeholders).[8] It was formulated in an inclusive and participatory manner, as reflected in the fact that public opinions were collected to revise the K-SDGs and indicators.

The most remarkable aspect was that the K-SDGs were drawn based on civil participation. Civil participation was accomplished using three methods: (1) direct participation (working group to design indicators; K-MGoS), (2) feedback participation (discussion panels, regional discussion panels, bulletin boards, online platforms for feedback collection), and (3) indirect participation for proliferation of the K-SDGs (perception survey, public ambassadors, exhibitions). A nationwide survey of the public to assess awareness of the SDGs was conducted twice in 2018. Survey respondents ranged from teenagers to elderly people in their 70s. As a result of the two surveys to assess awareness of sustainable development, over 80% of respondents answered that they felt it necessary to establish nationalized goals for sustainable development, laying the

[7] The participants in the joint working group were evenly distributed among the private sector, the public, and academia (total: 192; private: 87; public: 72; academia: 33). Specific targets had more private and public participants considering the characteristics of the goal, such as SDG 12 (responsible consumption and production) and SDG 13 (climate action).

[8] K-MGoS were composed of a total of 132 members, of which 90 were from institutions, and 42 from the general public. Institution participants were selected most often from the following sectors: NGOs (17 persons), disabled groups (17 persons), and regional communities (12 persons).

groundwork for vision-sharing with the global community. SDGs forums for comprehensive national debates were also held twice, in which the public, including the public and private joint expert working groups, directly presented their opinions on the target values and implementation plans for the detailed targets of K-SDGs prepared by relevant ministries. In the discussion, the specific contents of the K-SDGs, the process of establishment, and plans to reflect the opinions of diverse stakeholder groups were discussed. Following the debate, K-MGoS presentations were delivered by groups representing the disabled, NGOs, industry, animal ecosystems, women, local communities, youth, and migrants to express the opinions of their respective social groups (NCSD, 2019).[9]

3.2 Evolution of Government–Third Sector Partnership in South Korea

The natural expansion of civil participation that synched with the flow of policy change from the first to fourth basic plans for sustainable development was not accomplished in one stretch, nor was it a coincidence. As for South Korea, the symbiotic relationship between the government and civil society traces back to 1995, when the Korea International Cooperation Agency (KOICA) launched the Civil Society Cooperation Program. The program is a representative public–private partnership modality that incorporated private expertise into development projects with partners in the third sector including CSOs, social economy organizations, and academia. Although the main purpose of this partnership was to improve the quality of aid implementation in partner countries, it generated momentum for the third sector to establish and develop organizational capacities simultaneously with the development of global networks. An examination of how global networks for international development have transformed with the evolution of third sector participation levels is summarized in Table 1. Partnership levels can be divided into four phases depending on

[9] Adding both rounds, the discussion meeting gathered a total of 1,000 participants and public opinions were gathered from people of all social standings. Their opinions were used to establish the K-SDG detailed targets and indicator system, as well as to devise implementation plans for each relevant ministry. A total of 14 ambassadors were selected through direct voting by approximately 400,000 people, and public figures including actors, newscasters, and Olympic medalists were chosen to promote the K-SDGs.

the degree of their engagement with the government in the process of policy change.

The first three periods show relatively preliminary levels of third sector development prior to the adoption of the SDGs. Phase 1 refers to the informative level, at which the process of learning and spreading information occurs while putting more weight on searching for the right audience to transmit information. In South Korea, the enforcement of "globalization" by President Kim Young-Sam and his presidential statement made at the 1995 World Summit for Social Development in Copenhagen served as an unexpected catalyst to launch the aforementioned Civil Society Cooperation Program by KOICA, which marked the beginning of the partnership between the Korean government and civil society. With

Table 1 Third sector participation levels in line with policy change

Levels of third sector participation[10]	*Events and main junctures*	*Launch of global networks*	*Launch of civil society networks in Korea and policy change*
Phase 1: *Informative* level	• Agenda 21 (1992) • Copenhagen World Summit for Social Development (WSSD) (1995)	• CIVICUS (1993) • Reality of Aid Network (1993) • Social Watch (1995)	• KCOC (1999)
Phase 2: *Consultative* level	• Millennium Summit & Adoption of MDGs (2000)	• ActionAid International Federation (2003) • Global Call to Action Against Poverty (GCAP) (2003)	• GCAP Korea (2005) • ODA Watch (2006) • Strategic Plan for International Development Cooperation (2010) • *First Basic Plan for Sustainable Development* (2006–2011)

(continued)

[10] Labels for each level of participation from Phases 1 to 4 have been identified by Cázarez-Grageda (2018, 26).

Table 1 (continued)

Levels of third sector participation	*Events and main junctures*	*Launch of global networks*	*Launch of civil society networks in Korea and policy change*
Phase 3: *Empowering* level	• OECD High-Level Fora on Aid Effectiveness (2008 and 2011) and Istanbul Principles for CSO Development Effectiveness (2010)	• BetterAid & Open Forum for CSO Development Effectiveness (2010) • Beyond 2015 (2011) • CSO Partnership for Development Effectiveness (CPDE) (2012)	• KoFID (2010) • Beyond 2015 Korea (KCOC, KoFID, GCAP Korea) (2012) • Development Alliance Korea (DAK) (2012) • Framework Act on International Development Cooperation (2010) • Mid-term Strategy for Development Cooperation (2011–2015) • *Second Basic Plan for Sustainable Development* (2011–2015)
Phase 4: *Partnership* level	• 2030 Agenda for Sustainable Development and Adoption of SDGs (2015)	• Asia Development Alliance (ADA) (2013) • Action/2015 (2014) • Forus (former International Forum of National NGO Platforms [IFP/FIP]) (2018)	• Action/2015 Korea (KCOC, KoFID, GCAP Korea) (2015) • PIDA (former ODA Watch) (2016) • Mid-term Strategy for Development Cooperation (2016–2020) • *Third Basic Plan for Sustainable Development* (2016–2020)

(continued)

Table 1 (continued)

Levels of third sector participation	*Events and main junctures*	*Launch of global networks*	*Launch of civil society networks in Korea and policy change*
Phase 5: *All-inclusive participatory level with a whole-of-society approach*	• Creating an enabling environment for an all-inclusive and participatory partnership with the third sector in line with OECD's Policy Coherence for Sustainable Development		• Policy Framework for Government–Civil Society Partnership in International Development Cooperation (2019) • *Fourth Basic Plan for Sustainable Development* (2021–2040)

Source Authors

the launch of global networks such as CIVICUS and Social Watch, the Korea NGO Council for Overseas Development Cooperation (KCOC) was established in 1999 as a consultative group for development NGOs in South Korea engaged in humanitarian assistance and development cooperation projects.

Next, Phase 2 refers to the consultative level, at which non-state actors can be invited as a consultative counterpart to share expertise with the government on specific issues. Amidst implementation of the MDGs, gaining information through interaction with the third sector began to be considered crucial for facilitating development goals. During this period, the Global Call to Action Against Poverty (GCAP) and ActionAid were newly launched, and soon after came GCAP Korea that was established for advocacy functions and monitoring MDGs implementation by the Korean government. In the meantime, ODA Watch (2006) was founded with the aim of monitoring whether government work is proceeding with responsibility and accountability. Commemorating its 10th anniversary in 2016, ODA Watch underwent an innovative transformation by changing its network into the People's Initiative for Development Alternatives (PIDA).

During the empowering level of Phase 3, the CSO Partnership for Development Effectiveness (CPDE) was established as an international platform that merged BetterAid and Open Forum. The two were formerly operated separately prior to the HLF-4 in 2011. The CPDE represents civil society as a member of the executive committee of the Global Partnership for Effective Development Cooperation (GPEDC). In alliance with the global network called Beyond 2015, Beyond 2015 Korea was launched in 2012 by KCOC, GCAP Korea, and the Korea Civil Society Forum on International Development Cooperation (KoFID). The same year, Development Alliance Korea (DAK) was launched. DAK was the first public–private partnership among civil society, the private sector, academia, and diverse actors cooperating with the government on foreign aid. Throughout the first decade of the new millennium, non-state actors began to gain high recognition as an essential actor in development cooperation with emphasis placed on partnerships with civil society throughout the OECD High-Level Fora on Aid Effectiveness in Accra (2008) and Busan (2011). Signing of the Busan Partnership for Effective Development Cooperation was monumental as the Istanbul Principles for CSO Development Effectiveness were officially endorsed (Open Forum for CSO Development Effectiveness, 2010).

Basic forms of collaborative governance have been witnessed since 1995; however, the initial formality of the partnership in terms of domestic activities was primarily concentrated on the government's budget support of CSOs for advocacy purposes (KCOC, 2017). Meaningful participation and engagement in the policy process were exceptional, as third sector opinions were by and large taken in the form of co-optation of already-drafted policies that were regarded as final unless deficiencies and/or omissions were found to be problematic. In other words, coordination practices were mostly led by the government in a top-down approach without creating dynamism with the third sector. The turn from MDGs to SDGs called for a better understanding of combining traditional methods through top-down agendas initiated by the government and bottom-up approaches that guarantee roles for civil society as advocates, beyond being viewed as supporters (Kim, 2017).

The most recent stage refers to the Phase 4 partnership level, at which the 2030 Agenda for Sustainable Development has been proposed and adopted as SDGs since 2015. Having been built on the former three phases, the participation level of the third sector has reached the stage at which governments collaborate on equal terms to facilitate policies for specific issues. With SDG 17, which stipulates the goal to "strengthen the means of implementation and revitalize the global partnership for sustainable development," global networks stretched out further to encompass small and large coalitions worldwide. The former Beyond 2015 network was transformed into Action/2015 with a similar mission to form a consultative group among forums. Action/2015 Korea has greatly contributed to raising public awareness of the SDGs. Likewise, South Korea's civil society network has been well-aligned with the trajectories of global network development. The timely and coherent response to the changing atmosphere of the international development field has also contributed to South Korea's successful transition from a recipient to a donor. Moreover, constant demand for a comprehensive policy statement that defines the partnership modality between the government and civil society has been raised since 2017, with the OECD DAC peer review creating momentum.

3.3 Co-Optation to Inclusiveness with a Whole-of-Society Approach

As a response to the 2017 OECD DAC peer review, which recommended that the South Korean government strengthen partnerships with civil society (OECD, 2018, 22), the Ministry of Foreign Affairs actively began to search for means and communication channels to engage with the third sector. The pinnacle of efforts to seek policy coherence for sustainable development was witnessed with the establishment of the Policy Framework for Government–Civil Society Partnership in International Development Cooperation, enacted by the 32nd Committee for International Development Cooperation in January 2019 (CIDC, 2019). The institutional supplementation with the framework establishment and the systemic change by expanding the authorities and functions of the Committee for International Development Cooperation (CIDC) have opened doors toward a deepened partnership level between the third sector and the government. The incubating process takes time as the third sector expands its scope of participation from simply being supported and funded by the government; passive collaboration based on co-optation will eventually lead to a phase in which participation based on inclusive and equal partnership is nestled into the policy process.

In this context, the partnership between the government and civil society in South Korea is worthy of notice since it is on the verge of going beyond a Phase 4 partnership level into a more inclusive participatory level. Such systemic change is denoted as Phase 5 (participatory level with a whole-of-society approach in Table 1) and is a means to providing a blueprint showing the way forward for SDGs implementation with the principle of all-inclusiveness. Although not having reached its fullest scale yet, South Korea has been establishing sustainable development networks through which not only ministry-level but also CSO-level organizations partner on a sub-governmental platform to seek an enabling environment with a whole-of-society approach.[11] There is a total of 62 sustainable development networks under the NCSD: 28 organizations for the environment, five for energy, two for culture, one each for industry, ODA, the social dimension, technology, and land, and 20 global organizations that are in partnerships with the NCSD. Aside from the aforementioned areas, other sectors are covered by an expansive number of organizations operated either by government ministries, associations, networks, NPOs, CSOs, multilateral institutions, or research centers.[12] Collectively called sustainable development networks, the coordination ties may be looser than central government policies, but they still serve as grounds for stimulating policy coalitions among communities.

Moreover, UN Regional Centre of Expertise on Education for Sustainable Development (RCE) regional branches in South Korea are stationed in Tongyeong, Inje, Dobong-gu, and Changwon.[13] The main functions of the UN RCE are to develop international collaborative relationships to promote education in the Asia–Pacific region and develop education tools and programs to promote sustainable growth, among others. This aligns with the emphasis put on local government participation and representation from the third sector. Education in disperse regional areas of South Korea, and the joint partnership among the sub-national government and multinational institutions with a global agenda, highlights the fact that nongovernmental representatives can be middlemen for broadening the boundaries of reach. Also, under the UN ECOSOC NGO consultative body, the Local Sustainability Alliance of Korea (LSAK) is

[11] http://ncsd.go.kr/network?content=1.

[12] Full list of organizations can be found in Table 2 (Appendix).

[13] http://ncsd.go.kr/network?content=2.

supporting localization of the K-SDGs so that consultation and implementation of the SDGs can align with the local context as L-SDGs.[14] This requires greater involvement of stakeholders with a whole-of-society approach under multi-layers of networks that orchestrate coordination. The important stage to pass is the initial inception of such organizational structures formed through advocacy coalitions to create subsystems of networks, thereby inducing the government and civil society to meet on middle ground where neither hierarchical nor network governance rules to the fullest scale during transitional times.

4 Conclusion: Challenges Ahead

This chapter aimed not only to detect the trajectory of policy change in South Korea toward nationalizing and implementing the SDGs, but also to anticipate tasks ahead with regard to how an enabling environment for all-inclusive participation can be formed. To achieve effective development cooperation outcomes through nationalization of the SDGs, it was found that orchestrating a whole-of-society approach is the key to reaching the stage of all-inclusiveness, as shown in the case of South Korea. Implementation of the SDGs with a whole-of-society approach depends on how well the government recognizes its changing role from (i) a simple information provider and funder in the initial stage, toward (ii) a facilitator with brokerage functions among multi-stakeholders during the incubation stage, eventually transforming its role, as (iii) a participation enabler through the establishment of legal and policy frameworks that guarantee civil society engagement and autonomous activities by stakeholders as equal participants. Indeed, transition between participation levels does not occur in a clear-cut manner. Shifts in global paradigms do not automatically lead to moving all states from one stage to another. Instead, as partnerships proceed from one phase to the next, the strength of each builds cumulatively on top of the former, which involves maintaining the features of the former while expanding the scope of participation.

[14] http://www.sdkorea.org/contents/business/business_03.php.

SDGs are now considered a norm within international society beyond simply the domain of development cooperation. However, the level of achievement in each country depends on its implementation capacity and voluntary efforts to align national policies with global development goals. As an interim review, states are strongly advised to share their SDGs implementation progress report with the UN High-Level Political Forum. Often, such external pressure can only function as normative grounds for reducing domestic resistance against policy change instead of a stimulus that accelerates a true means of implementation (McAdam et al., 2001). Internally, cooperation among implementing agencies is a prerequisite to gain a sufficient level of policy coherence. As SDG target 17.14 specifies, the coherence of country policies with sustainable development objectives should be improved. Thus, it is crucial to understand the current policy coherence status of Korea to further enhance public awareness and to carry out coordinated actions with a whole-of-society approach. This calls for well-tuned collaboration among multi-stakeholders including government agencies, CSOs, and private sector entities (OECD, 2016b).

However, since the inception of the SDGs in 2015, development cooperation in South Korea has been criticized in terms of a lack of public awareness of the SDGs, the fragmented structure of the implementation processes for official development assistance (ODA), and the domination of developmentalism over the SDGs. To address the issues at stake in terms of development cooperation policies for SDGs implementation in Korea, this chapter examined the OECD's PCSD with ACF. The advocacy coalition framework contributes to logically illuminating complex policy changes such as policy conflicts, but there is a limit to addressing the clear starting point of policy change and locating where interaction occurs during policy learning (Matti & Sanstrom, 2011). There are obvious limitations in that the temporal validity of the framework normally lasts for only about a decade, but considering that SDGs implementation is aimed to be accomplished by 2030, the short-term framework for approaching the implementation mechanism can be feasible in the current phase in which South Korea stands between partnership–participatory levels. The framework indicates that under systemic external conditions of a globally agreed norm of sustainable development, policy-oriented learning with advocacy coalitions amid the policy process is fundamental to forming an enabling environment for SDGs implementation in South Korea.

To incorporate the SDGs into national policy-making, there is a variety of methods to devise coordination mechanisms by improving existing bodies or creating new ones: for example, a new ministry-level hub, high-level inter-ministerial commission, or a national council on sustainable development (NCSD). The primary responsibilities of such a group should include developing and coordinating SDGs implementation strategies, developing a national monitoring framework and an accompanying set of national indicators, consulting with key stakeholders, and preparing for regional and global dialogues on SDGs implementation. Thus, the chapter attempted to trace the evolution of advocacy coalitions in South Korea by providing an overview of the levels of civil society participation in accordance with the development of global networks. It then deduced that the forthcoming phase should accompany participation from all levels with legal safeguards and policy processes that engage civil society. Lack of coordination strategies for SDGs implementation can therefore be complemented by forming a partnership-enabling environment with a whole-of-society approach to Korea's policy process, which can lead to systemic change that improves government–third sector relationships from a form of co-optation towards inclusiveness.

Appendix

See Table 2.

Table 2 Sustainable development networks for all-inclusiveness

Classification	*Operation*	*Organization*
Environment	Ministry of Environment	Metropolitan Air Quality Administration Living Environment Safety Information System-Green Nuri ABSCH Genetic Resources Information Center I Love Water
	Ecoclub	Ecoclub
	Environment Action Association	Environment Action Association
	Green Fund	Green Fund
	Green Korea	Green Korea
	Environmental Justice	Environmental Justice
	Environmental Preservation Association	Environmental Preservation Association
	Korea Zero Waste Movement Network	Korea Zero Waste Movement Network
	National Council of the Green Consumers Network in Korea	National Council of the Green Consumers Network in Korea
	Korea Federation for Environmental Movements (KFEM)	Korea Federation for Environmental Movements (KFEM)
	The Climate Reality Project	The Climate Reality Project
	Green Education Center	Green Education Center
	Greenpeace	Greenpeace Korea
	Forest for Life	Forest for Life
	National Environmental Education Center	National Environmental Education Project
	Korea Environment Corporation	Korea Environment Corporation Climate Change Promotion Portal
	National Institute of Biological Resources	National Institute of Biological Resources
	Greenhouse Gas Inventory and Research Center, Ministry of Environment,	Greenhouse Gas Inventory and Research Center, Ministry of Environment
	Korea Meteorological Administration	Korea Meteorological Administration Global Atmosphere Watch
	National Institute of Environmental Research	National Institute of Environmental Research

(continued)

Table 2 (continued)

Classification	*Operation*	*Organization*
	National Biodiversity Center	National Biological Resource Management System
	Korea Environment Institute	Korea Environment Institute
	Korea Climate & Environment Network	Korea Climate & Environment Network
	Korea Association for UN Environment	Korea Association for UN Environment
Industry	Korea Business Council for Sustainable Development	Korea Business Council for Sustainable Development
Energy	Energy Peace (NPO)	Energy Peace (NPO)
	Korea NGO's Energy Network	Korea NGO's Energy Network
	Seoul Metropolitan Government	Seoul Metropolitan Government Eco Mileage
	Seoul Energy Dream Center	Seoul Energy Dream Center
	Korea Energy Agency	Korea Energy Agency
ODA	KOICA	KOICA ODA Information Portal
Social Dimension	Local Governments for Sustainability (ICLEI)	ICLEI Korea
Culture	The National Trust of Korea	The National Trust of Korea
	Korean National Commission for UNESCO	Korean National Commission for UNESCO
Technology	Green Technology Center	Green Technology Center
Land	Land Portal	Land Portal—SDGs
Global Organizations	WHO	WHO—SDGs
	The Ramsar Convention Secretariat	The Ramsar Convention Secretariat
	WNO	World Meteorological Organization
	UN	UN Sustainable Development Goals Knowledge Platform
		United Nations University, Institute for the Advanced Study of Sustainability (UNU-IAS)
	UNDP	United Nations Development Programme (UNDP)
	SDSN	Sustainable Development Solutions Network (SDSN)

(continued)

Table 2 (continued)

Classification	*Operation*	*Organization*
	FAO	FAO
	WRI	World Resources Institute (WRI)
	ESCAP	UN ESCAP
	World Bank	World Bank
	IISD	International Institute for Sustainable development (IISD)
	GEF	Global Environment Facility (GEF)
	IPCC	The Intergovernmental Panel on Climate Change (IPCC)
	OECD	OECD Climate Change
	UN FCCC	United Nations Framework Convention on Climate Change (UNFCCC)
	OUR PLANET	OUR PLANET
	IEA	International Energy Agency (IEA)
	GGGI	Global Green Growth Institute (GGGI)
	ICE	Chicago Climate Exchange (CCX)
Undefined	Korea Center for Sustainable Development (KCSD)	Korea Center for Sustainable Development (KCSD)
	Local Sustainability Alliance of Korea	Local Sustainability Alliance of Korea

Source http://ncsd.go.kr/network

References

Arstein, S. (1969). A ladder of citizen participation. *Journal of the American Planning Association, 4*(35), 216–224.

Baumgartner, F., & Jones, B. (1993). *Agendas and instability in American politics*. Chicago University Press.

Beck, U., Bonss, W., & Lau, C. (2003). The theory of reflexive modernization: Problematic, hypotheses and research programme. *Theory, Culture & Society, 20*(2), 1–33.

Beetham, D. (1996). *Bureaucracy*. Open University Press.

Bennett, C., & Howlett, M. (1992). The lessons of learning: Reconciling theories of policy learning and policy change. *Policy Sciences, 25*, 275–294.

Cázarez-Grageda, K. (2018, October). *The Whole of Society Approach: Levels of Engagement and Meaningful Participation of Different Stakeholders in the review process of the 2030 Agenda* (P4R Discussion Paper).

Cerna, L. (2013). *The nature of policy change and implementation: A review of different theoretical approaches*. OECD Publishing.
Chandler, D. (2004). *Constructing global civil society*. Palgrave Macmillan.
Chisholm, D. (1989). *Coordination without hierarchy: Informal structure in multiorganizational systems*. University of California Press.
CIDC. (2019, January 15). *Policy Framework for Government–Civil Society Partnership in International Development Cooperation* (Resolution 32–3).
CIVICUS. (2019). *State of civil society report 2019: The year in review*. CIVICUS. https://www.civicus.org/documents/reports-and-publications/SOCS/2019/state-of-civil-society-report-2019_executive-summary.pdf. Accessed on 7 February 2021.
Common, R. (2013). When policy diffusion does not lead to policy transfer: Explaining resistance to international learning in public management reform. In P. Carroll & R. Common (Eds.), *Policy transfer and learning in public policy and management: International contexts, content and development* (pp. 13–29). Routledge.
GPEDC. (2016). *Nairobi Outcome Document*. https://www.effectivecooperation.org/system/files/2020-05/Nairobi-Outcome-Document-English.pdf. Accessed on 7 February 2021.
Haas, P. M. (1989). Do regions matter? Epistemic communities and Mediterranean pollution control. *International Organization, 43*(3), 377–403.
Hardin, G. (1968). The tragedy of the commons. *Science, 162*(3859), 1243–1248.
Hedman, J., & McDonnell, I. (2011). *How DAC members work with civil society organisations: An overview*. OECD Publishing.
Jessop, B. (1997). The governance of complexity and the complexity of governance: Preliminary remarks on SOME problems and limits of economic guidance. In A. Amin & J. Hausner (Eds.), *Beyond Market and Hierarchy: Interactive Governance and Social Complexity* (pp. 95–128). Edward Elgar.
Jessop, B. (2003). Governance and meta-governance failure: On reflexivity, requite variety and requisite irony. In H. Bang (Ed.), *Governance as Social and Political Communication* (pp. 101–116). Manchester University Press.
KCOC. (2017, April 24). *OECD DAC Peer Review Korean Civil Society Report*. KCOC.
Keck, M. E., & Sikkink, K. (1998). *Activities beyond borders-advocacy networks in international politics*. Cornell University Press.
Kim, T. K. (2017). From supporters to advocates: A study on the identity transition of development civil society organizations in South Korea. *Discourse 201, 20*(3): 111–155.
Kindornay, S., & Zeki, K. (2019). *A Whole-of-Society Approach: Partnerships to realize the 2030 Agenda Good Practice in 2030 Agenda Implementation*

Series. British Columbia Council for International Cooperation and Canadian Council for International Cooperation.

Kuenkel, P. (2017). A pattern approach to sustainability transformation-how the 17 SDGs can become a starting point for systemic change. *Collective Leadership Studies, 5*, 2569–1171.

Lewis, D. (2001). *The management of non-governmental development organizations: An introduction*. Routledge.

Lewis, D. (2014). *Non-governmental organizations, management and development*. Routledge.

Lijphart, A. (1999). *Patterns of democracy*. Yale University Press.

Lindblom, C. E. (1977). *Politics and markets*. Basic.

Mahoney, J. (2000). Path dependence in historical sociology. *Theory and Society, 29*(4), 507–548.

Matti, S., & Sanstrom, A. (2011). The rationale determining advocacy coalitions: Examining coordination networks and corresponding beliefs. *Policy Studies Journal, 39*(3), 385–410.

McAdam, D., Tarrow, S., & Tilly, C. (2001). *Dynamics of contention*. Cambridge University Press.

Meadowcroft, J. (2010). Sustainable development. In M. Bevir (Ed.), *The SAGE handbook of governance* (pp. 535–551). Sage.

Meuleman, L. (2018). *Metagovernance for sustainability: A framework for implementing the sustainable development goals*. Routledge.

Monkelbaan, J. (2019). *Governance for the sustainable development goals*. Springer.

National Commission on Sustainable Development (NCSD). (2019). *A Report on Korean Sustainable Development Goals (K-SDGs)*. Ministry of Environment, South Korea. http://ncsd.go.kr/api/1572586270021_K-SDGs_report_eng.pdf. Accessed on 3 May 2021.

OECD. (2009). *Civil society and aid effectiveness: Findings, recommendations and good practice*. OECD Publishing.

OECD. (2011, November 29–December 1). *Busan Partnership for Effective Development Co-operation, Fourth High-Level Forum on Aid Effectiveness*. Busan. www.oecd.org/dac/effectiveness/49650173.pdf. Accessed on 7 February 2021.

OECD. (2012a). *OECD development co-operation peer reviews: Korea 2012*. OECD Publishing.

OECD. (2012b). *Partnering with civil society: 12 lessons from DAC peer reviews*. OECD Publishing.

OECD. (2016a). *Better policies for sustainable development 2016: A new framework for policy coherence*. OECD Publishing.

OECD. (2016b). *Private sector engagement for sustainable development: Lessons from the DAC*. OECD Publishing.

OECD. (2018). *OECD development co-operation peer reviews: Korea 2018*. OECD Publishing.
OECD. (2019). *Policy coherence for sustainable development 2019: Empowering people and ensuring inclusiveness and equality*. OECD Publishing.
Offe, C. (2009). Governance: An empty signifier? *Constellations, 16*(4), 550–562.
Open Forum for CSO Development Effectiveness. (2010, September 28–30). *Istanbul CSO Development Effectiveness Principles, endorsed at the Open Forum's Global Assembly*. Istanbul.
Ostrom, E. (1990). *Governing the commons*. Cambridge University Press.
P4R. (2018). *Interviews: The Whole of Society Approach: Different Levels of Engagement and Meaningful Participation*.
Pierson, P. (2000). Increasing returns, path dependence, and the study of politics. *The American Political Science Review, 94*(2), 251–267.
Rhodes, R. A. W. (1997). *Understanding governance: Policy networks, governance, reflexivity and accountability*. Open University Press.
Rhodes, R. A. W. (2000). Governance and public administration. In J. Pierre (Ed.), *Debating governance: Authority, steering and democracy* (pp. 54–90). Oxford University Press.
Rhodes, R. A. W. (2007). Understanding governance: Ten years on. *Organization Studies, 28*(8), 1243–1264.
Sabatier, P. (1986). Top-down and bottom-up approaches to implementation research: A critical analysis and suggested synthesis. *Journal of Public Policy, 6*(1), 21–48.
Sabatier, P. (1988). An advocacy coalition framework of policy change and the role of policy learning therein. *Policy Sciences, 21*, 128–168.
Sabatier, P., & Jenkins-Smith, H. (1993). Policy change over a decade or more. In A. Sabatier & H. Jenkins-Smith (Eds.), *Policy change and policy learning: An advocacy coalition approach* (pp. 13–39). Westview Press.
Sabatier, P., & Jenkins-Smith, H. (1999). The advocacy coalition framework: An assessment. In A. Sabatier (Ed.), *Theories of the policy process* (pp. 117–166). Westview Press.
Scharpf, F. W. (1997). *Games real actors play: Actor-centered institutionalism in policy research*. Westview Press.
Scholte, J. (2011). Conclusion. In J. Scholte (Ed.), *Building global democracy? Civil society and accountable global governance* (pp. 306–342). Cambridge University Press.
Seery, E., & Seghers, J. (2019, April 7). *Hitting the target: An agenda for aid in times of extreme inequality* (Oxfam Briefing Paper). https://reliefweb.int/sites/reliefweb.int/files/resources/bp-hitting-the-target-aid-inequality-agenda-080419-en.pdf. Accessed on 7 February 2021.

Sørensen, E., & Torfing, J. (2005). The Democratic anchorage of governance networks. *Scandinavian Political Studies, 28*(3), 195–218.

Torfing, J., Peters, B. G., Pierre, J., & Sørensen, E. (2012). *Interactive governance: Advancing the paradigm*. Oxford University Press.

United Nations Committee of Experts on Public Administration. (2018, April 23–27). *Report on the seventeenth session* (Economic and Social Council, Official Records, 2018; Supplement No. 24). E/2018/44-E/C.16/2018/8. https://undocs.org/en/e/c.16/2018/8. Accessed on 7 February 2021.

United Nations Committee of Experts on Public Administration. (2019, April 8–12). *Progress on Institutional Aspects of Sustainable Development Goal 16: Access to information, transparency, participation, and accountability* (Economic and Social Council, Official Records, 2019. E/C.16/2019/7). https://undocs.org/E/C.16/2019/7. Accessed on 7 February 2021.

Woo, C. B., Lee, J. H., & Kim, T. K. (2019). Strategy of governance and public administration for implementing Sustainable Development Goals (SDGs): Focusing on Korea's development cooperation policies. *Journal of Governance Studies, 14*(2), 269–298.

Wood, J., & Fällman, K. (2019). *Enabling Civil Society: Select survey findings* (OECD Development Co-operation Working Papers, No. 57). OECD Publishing.

Websites

Korean Law Information Center. Basic Act for Sustainable Development. https://law.go.kr/LSW/lsInfoP.do?lsiSeq=80019#0000. Accessed on 3 May 2021.

Korean Law Information Center. Sustainable Development Act. https://www.law.go.kr/%EB%B2%95%EB%A0%B9/%EC%A7%80%EC%86%8D%EA%B0%80%EB%8A%A5%EB%B0%9C%EC%A0%84%EB%B2%95. Accessed on 3 May 2021.

Korean Law Information Center. Basic Act on Low Carbon Green Growth. https://www.law.go.kr/%EB%B2%95%EB%A0%B9/%EC%A0%80%ED%83%84%EC%86%8C%EB%85%B9%EC%83%89%EC%84%B1%EC%9E%A5%EA%B8%B0%EB%B3%B8%EB%B2%95. Accessed on 3 May 2021.

Ministry of Environment of South Korea. Sustainable Development Portal. http://ncsd.go.kr/. Accessed on 3 May 2021.

Ministry of Environment of South Korea. A Report on Korean Sustainable Development Goals (K-SDGs). http://ncsd.go.kr/nationaleffort?content=1. Accessed on 3 May 2021.

Conclusion

Huck-ju Kwon, Tatsufumi Yamagata, Eunju Kim and Hisahiro Kondoh

This book is made in order to explore the evolution of foreign aid policy in Japan and South Korea and to seek new strategies in an uncertain world. The world in the 2020s is becoming more unpredictable than in the previous decade. COVID-19 broke out at the very end of the 2010s and reached every corner of the globe by 2020. Even with some effective vaccines at hand, many parts of the world are suffering from occasional lockdowns up to mid-2021 (Kwon, 2020). The switch of presidency in the United States from Donald Trump to Joseph Biden was another factor dramatically altering the landscape of global politics. The Trump administration symbolized the unilateralism pursued by some world leaders (Kondoh, 2019). President Biden converted US policies for international cooperation and restored multilateralism to collaborate with allied states. Confronted with the transforming United States, China and Russia have strengthened their presence amid the COVID-19 pandemic. Armed conflicts taking place in Africa, Asia, and the Middle East are amplifying the uncertainty in the world economy and politics.

Under the unforeseeable circumstances of the world, the roles of Japan and South Korea, which are two developed countries in East Asia, have to be revisited and redefined. For that purpose, this concluding chapter firstly summarizes what each pair of chapters discussed in the main body of this book in section "Japan and South Korea: Similarities and Contrasts".

H.-j. Kwon et al. (eds.), *International Development Cooperation of Japan and South Korea*,
https://doi.org/10.1007/978-981-16-4601-0

It is impressive that there are some common features between Japan and South Korea, some of which stem from history and geographical proximity. In the meantime, there are some contrasts in evolution and progress between the two countries toward becoming East Asian donors. Those similarities and contrasts will be reviewed to elucidate the standpoints of the two countries and to figure out common new strategies ahead.

Finally, section "New Strategies for International Development Cooperation" attempts to address the questions posed in the Introduction that are examined throughout this book, which relate to (i) the dynamics of policy making in Japanese and Korean ODA, (ii) relative weights distributed between national interests and the creation of the global public good, and (iii) new ODA strategies hereafter. Answers to these questions are developed into policy recommendations to the governments and the general public of the two countries.

Japan and South Korea: Similarities and Contrasts

This book is organized with pairs of chapters which describe key issues in Japan and South Korea. Naturally, similarities and differences between the two countries arise from the comparison between companion chapters. These similarities and contrasts themselves and the factors generating the similarities and contrasts are insightful to figure out new strategies of international development cooperation for Japan and South Korea in the 2020s.

Experiences of Industrial Development as Aid Recipients

There are striking similarities in features of ODA between Japan and South Korea, as detailed in several chapters of this book. Below, these similarities are discussed, followed by contrasts.

A common feature from which other commonalities between the two countries originate is the experience of industrial development as aid recipients from war devastation. Most cities in Japan were destroyed by bombing at the end of World War II in 1944–1945. Obtaining financial assistance mainly from the United States, Japan rebuilt the country. Eventually, Japan recovered diplomatic relations with other countries in 1952 and started industrialization. War reparation was a condition of return to the international community as well as the origin of Japan's aid to neighboring countries (Chapter 3 by Sato, 2022). South Korea was also severely damaged by the Korean War. With overseas assistance, South Korea struggled through reconstruction and industrial development to become an OECD member country in 1996. As Kwon (2022) reviewed in Chapter 2 of this volume, the achievement of industrial development is a point of national pride for the Korean people, which created momentum for the Knowledge Sharing Program (KSP) of the country.

Emphasis on Economic Cooperation for Industrial Development in Recipient Countries

The experiences of industrial development with foreign assistance of both Japan and South Korea were translated into their stances on international development cooperation for developing countries. Both countries reflected on their own patterns of industrial development with assistance to construct aid policies toward neighboring countries. Neighboring East Asian countries, which have been the main recipients of aid from Japan and South Korea, had made considerable progresses in economic development by the end of last century. Hence, their demands for cooperation also accorded with the orientation of Japan and South Korea toward assistance emphasizing the economic development of recipient countries, infrastructure building with preferential loans, and the involvement of the private sector. Several chapters in this book (Chapters 3–6) document these features as shared by the two countries. In fact, many of these features are applicable to China, too (Asplund & Söderberg, 2017; Shimomura & Ping, 2013; Stallings & Kim, 2016).

It is notable that the features are mutually reinforcing and institutionally complementary. The emphasis on economic infrastructure is realized by the availability of preferential loans. Prospects for economic infrastructure building might allow the private sector to find opportunities for deeper involvement in ODA. Above all, the image of economic prosperity due to assistance from foreign aid from the point of view of recipient countries is shared by those engaged in both the public and private sectors in Japan and South Korea through their own experiences with development. Thus, as detailed in Chapters 7 and 8, the mobilization of the private sector and the advanced technology developed in the two countries are considered conducive to the further development of East Asian recipient countries.

Reactive Rationale Setting

Since, the model of international development cooperation was formed through experiential success with industrial development, Japan and South Korea lagged behind in the conceptualization of normative rationales for their international development cooperation. Criticism and pressure from the OECD/DAC and the United States to follow the norms and principles set by forerunners were the main driving forces for Japan to catch up with other member countries of the OECD/DAC. Public criticism regarding inefficiency and corruption in ODA projects also created momentum to develop the philosophy of Japan's foreign aid (Chapters 1 and 3).

South Korea's international development cooperation originated from the motivation to compete with North Korea for recognition of the country in the Free World (Chapter 4 by Song & Kim, 2022). After South Korea was admitted as a newly industrializing economy (NIE) in the 1980s, growing trade surplus urged to the country to allow the outflow of foreign exchanges in the form of assistance. Thus, diplomatic and economic situations preceded the formation of rationales in South Korea.

Efforts to Balance Universal Values and National Interests

A natural consequence of the former two features is a struggle to reconcile global norms stressed by western donors and the economic-growth-oriented aid model of Japan and South Korea. Kim (2022) depicts this struggle for South Korea as "efforts to balance universal values and national interests" in Chapter 6, while Jung and Takahashi (2022) analyze the same endeavor for Japan in Chapter 5.

Some of the public in the two countries simply assume that any government expenditure, even ODA, must be spent for the direct benefit of nationals in each country. From this viewpoint, the reflection of national interests in each country's ODA sounds natural. However, sustaining universal values is the duty of an established member of the global community. Failing to uphold this obligation affects the status of the two countries. Japan and South Korea face this dilemma more strongly than western donors, partly because both countries are recognized as relative newcomers to the society of donors and partly because the image of destitution due to war followed by economic development with foreign aid is heavily imprinted on older generations in the two countries.

Weak Partnership with Civil Society

A consequence of placing greater weight on pursuing national interests within the private sector was a weak partnership with civil society in both countries. Statistics clearly show this small contribution of the governments to civil society. Takayanagi (2022) stated in Chapter 9 that while DAC members on average allocated around 15% of their bilateral official development assistance to CSOs in 2018, Japan and South Korea allocated only 1.7% and 2.1%, respectively.

The government of Japan initiated a funding scheme to CSOs in 1989. Since then, having experienced the Kobe Earthquake of 1995 and Great East Japan Earthquake of 2011, the government has gradually enhanced recognition of the roles of CSOs in overcoming both domestic and oversea challenges. South Korean CSOs were mainly engaged in service delivery for ODA in the early 1990s. They have now formed the Korea Civil Society Forum on International Development Cooperation (KoFID) and monitor Korean ODA policy to strengthen the advocacy role (Chapters 10 and 12).

Notable Differences

After reviewing the important similarities described above, differences between Japan and South Korea look marginal. As Japan joined the OECD/DAC around five decades before South Korea joined in 2010, Japan's ODA policy was a close reference for South Korea in order to formulate the framework of its international development cooperation (Kondoh, 2013). However, that became another factor producing the similarities raised above.

Some differences still remain. As an institutional aspect, South Korea formulated the Framework Act on International Development Cooperation, which oversees the ODA of the country, in 2009. Japan has not enacted any law covering its international development cooperation. Only some charters on ODA were approved and revised twice by the Cabinet. In the meantime, Japan integrated two aid agencies, the Japan International Cooperation Agency (which formerly took charge of technical assistance) and (part of) the Japan Bank for International Cooperation (which administered ODA loans), into a single aid agency in 2008. South Korea maintains two counterpart aid agencies: the Korea International Cooperation Agency and the Economic Development Cooperation Fund.

A contemporary issue exhibiting a contrast is national implementation of the SDGs. Japan considers the SDGs to be country-wise and comprehensive goals from which Japan can also benefit. The Prime Minister and a nationwide business association, Keidanren, handle SDGs for promotion in Japan (Chapter 11). South Korea highlights some features of the SDGs by customizing the goals as the Korean Sustainable Development Goals (K-SDGs). The Ministry of Environment mainly takes charge of SDG issues on behalf of the country. Science, technology, and innovation (STI) may have a potential role in achieving the K-SDGs (Chapters 8 and 12).

New Strategies for International Development Cooperation

In the beginning of the third decade of this millennium, we all live in an uncertain world. COVID-19, climate changes, rivalry between superpowers, regional armed conflicts, and subsequent forced migration are sources of uncertainty and concern. However, it does not seem that the citizens of the world are sufficiently united to solve these critical issues. There are conflicts and confrontations among countries, social strata, ethnic groups, and religious groups.

The contributors to this book, who are all Korean and Japanese, aspire to encourage Japan and South Korea to play constructive roles in reducing uncertainty in the world and helping developing countries make themselves prosperous. To do so, what will be our new strategies for international development cooperation? In this concluding chapter, some proposals for new strategies are given.

The following three questions were posed in the Introduction. (1) What are the underlying dynamics of policy making in Japanese and South Korean ODA? (2) Will national interests be the goal of international development cooperation, or will creating the global public good be the main objective of these countries' ODA policies? (3) Do Japan and South Korea have new ODA strategies for an uncertain world? The chapters in the main body of this book incorporate answers to the first two questions.

The answers to questions (1) and (2) then lead to some proposals which address question (3).

Global Interests to Be Prioritized for International Development Cooperation

As an answer to question (1), salient features of the dynamics of ODA policy making in Japan and South Korea are the mobilization of the private sector and the reconciliation of efforts to achieve universal values, such as poverty reduction and sustainable development, and national interests. Even though many donors encounter the same reconciliation issue in harmonizing the pursuit of public interests and the appeal for contributions from a particular donor, Japan and South Korea encounter stronger forces in terms of national interests and economic cooperation.

However, national interests cannot be a main *raison d'etre* behind the international development cooperation of Japan and South Korea. When a donor says "this project is a win–win for both the donor and the recipient", the recipient may doubt whether or not the donor's "win" is greater than the recipient's "win". Because of the feature emphasizing economic development of the recipient and involvement of the private sector in Japan and South Korea, ODA of the two countries is more likely to be regarded as national-interest-oriented. For the two countries to be respected due to their international contributions, the principle the global public good outweighing the achievement of national interests must be strictly maintained.

Enlarging the Scope of the Private Sector

The feature of deep involvement by domestic private firms in Japan and South Korea creates a risk that the scope and potential of domestic firms define the domain of ODA for Japan and South Korea, which is particularly applicable to public–private partnership (PPP) projects. It is noticeable that there are opposite cases in which a wider scope of global firms expands the potential of ODA. A well-known example is the contribution of the Bill and Melinda Gates Foundation to global health. However, PPP projects of Japan and South Korea are likely to limit partner firms to only those from each country. In this case, Japanese and South Korean partner firms of PPP tend to replicate or reproduce activities based on which they already established advantages in the past.

This means that Japanese and South Korean firms want donor agencies to accommodate partner firms' original activities. If this is the case, even though some call this is a "win–win project", a "win" awarded to the Japanese or South Korean partner firm would be considerably greater than the other "win" granted to beneficiaries in recipient countries.

To avoid such unbalanced win–win projects, Japan and South Korea have to make efforts to enlarge the scope of domestic partner firms. PPP projects should be opportunities for Japanese and Korean firms to expand their capacity, which will generate more benefits for recipient countries. Meeting new demands and preferences in recipient countries would inspire motivation and capability in partner firms to accommodate them. Through guidance and encouragement by donor agencies of Japan and South Korea to partner firms, true win–win projects which benefit recipient countries and partner firms alike could be achieved.

Synergy of Collaboration of the Two Similar Countries

Generally speaking, a collective action by two similar actors intensifies its impacts. By leveraging a common strength of the two actors, a greater effect can be produced.

Collaboration between Japan and South Korea could generate such synergy if the two countries were well coordinated. Suppose, there is a recipient country which violates human rights and that Japan and South Korea agreed with the country to provide ODA for two different infrastructure projects. If Japan and South Korea jointly express concern for human rights protection and indicate the withdrawal of the two infrastructure projects, the impacts of the collective action on the recipient country will be magnified.

Two similar countries might compete with each other by featuring common strengths. This is where the rivalry with China regarding infrastructure building originates. As Asian donors which want to act as good neighbors in East Asia, Japan and South Korea as well as China will compete to attract countries in the region. ODA will be used as an instrument to do so. However, an essential part of international development cooperation is simple compassion applied to anyone and everyone in the world. This book also originates from the spirit of mutual cooperation between two groups of scholars in Japan and South Korea. This spirit of mutual cooperation and respect may go beyond the political tension between the two countries to reach neighboring states and even remote countries in order to build the global public good together with everyone on the globe.

References

Asplund, A., & Söderberg, M. (2017). An Asian aid paradigm: Japan leading from behind. In A. Asplund & M. Söderberg (Eds.), *Japanese development cooperation: The making of an aid architecture pivoting to Asia* (pp. 197–207). Routledge.

Jung, H., & Takahashi, M. (2022). Quest for sublation of economic development and poverty reduction: Dual features of Japan's aid in the post-cold war era and after. In H.-J. Kwon et al. (Eds.), *International development cooperation of Japan and South Korea: New strategies for an uncertain world* (pp. 105–128). Palgrave Macmillan.

Kim, E. (2022). Balancing universal values and economic interest through development cooperation in Korea. In H.-J. Kwon et al. (Eds.), *International development cooperation of Japan and South Korea: New strategies for an uncertain world* (pp. 129–148). Palgrave Macmillan.

Kim, S. G., & Hong, J. (2022). The government-civil society relationship in Korean international development cooperation from a historical perspective. In H.-J. Kwon et al. (Eds.), *International development cooperation of Japan and South Korea: New strategies for an uncertain world* (pp. 237–257). Palgrave Macmillan.

Kim, T. K., & Kim, B. K. (2022). A way forward to achieve the SDGs in Korea: Reformulating an enabling environment for sustainable development. In H.-J. Kwon et al. (Eds.), *International development cooperation of Japan and South Korea: New strategies for an uncertain world* (pp. 283–311). Palgrave Macmillan.

Kondoh, H. (2013). Korea's pathway from recipient to donor: How does Japan matter? In J. Sato & Y. Shimomura (Eds.), *The rise of development donors in Asia: Japan's impact on the evolution of emerging donors* (pp. 133–154). Routledge.

Kondoh, H. (2019). Amid the rise of unilateralism: Reinventing Multilateral cooperation and roles of Northeast Asian countries to achieve the SDGs. *Journal of International Development Studies, 28*(3), 1–3.

Kwon, H.-J. (2020). Policy options for international development cooperation in the COVID pandemic: A global perspective. *International Development and Cooperation Review, 12*(3), 23–29.

Kwon, H.-J. (2022). Reflection on a normative rationale for Korean ODA policy: Duty, self-regards, and obligation. In H.-J. Kwon et al. (Eds.), *International development cooperation of Japan and South Korea: New strategies for an uncertain world* (pp. 23–43). Palgrave Macmillan.

Ohno, I., & Uesu, S. (2022). New partnership with the private sector in Japanese development cooperation. In H.-J. Kwon et al. (Eds.), *International development cooperation of Japan and South Korea: New strategies for an uncertain world* (pp. 149–178). Palgrave Macmillan.

Park, K. R. (2022). Science, technology and innovation in sustainable development cooperation: Theories and practices in South Korea. In H.-J. Kwon et al. (Eds.), *International development cooperation of Japan and South Korea: New strategies for an uncertain world* (pp. 179–208). Palgrave Macmillan.

Sato, Jin (2022). Pivotal moments in Japanese ODA: Circa 1950–2010. In H.-J. Kwon et al. (Eds.), *International development cooperation of Japan and South Korea: New strategies for an uncertain world* (pp. 45–72). Palgrave Macmillan.

Shimomura, Y., & Ping, W. (2013). The evolution of "Aid, Investment, Trade Synthesis" in China and Japan. In J. Sato & Y. Shimomura (Eds.), *The rise of Asian donors: Japan's impact on the evolution of emerging donors* (pp. 114–132). Routledge.

Song, J.-S., & Kim, E. M. (2022). Institutional development of development cooperation and partnership in South Korea: History, institutions and quantitative overview. In H.-J. Kwon et al. (Eds.), *International development cooperation of Japan and South Korea: New strategies for an uncertain world* (pp. 73–101). Palgrave Macmillan.

Stallings, B., & Kim, E. M. (2016). Japan, Korea, and China: Styles of ODA in East Asia. In H. Kato, J. Page, & Y. Shimomura (Eds.), *Japan's development assistance: Foreign Aid and the post-2015 agenda* (pp. 120–134). Palgrave Macmillan.

Takayanagi, A. (2022). Civil society organizations as partners and critics of Japan's Aid policy. In H.-J. Kwon et al. (Eds.), *International development cooperation of Japan and South Korea: New strategies for an uncertain world* (pp. 211–236). Palgrave Macmillan.

Takeuchi, S. (2022). Policy concepts and normative rationales in Japan's Foreign Aid: Human security, TICAD, and free and open Indo-Pacific. In H.-J. Kwon et al. (Eds.), *International development cooperation of Japan and South Korea: New strategies for an uncertain world* (pp. 3–22). Palgrave Macmillan.

Yamagata, T. (2022). Japan's approach to the SDGs: Decoupling between the SDGs and International Development. In H.-J. Kwon et al. (Eds.), *International development cooperation of Japan and South Korea: New strategies for an uncertain world* (pp. 259–282). Palgrave Macmillan.

Index

H.-j. Kwon et al. (eds.), *International Development Cooperation of Japan and South Korea*,
https://doi.org/10.1007/978-981-16-4601-0

M

N

The manufacturer's authorised representative in the EU is Springer
Nature Customer Service Centre GmbH, Europaplatz 3, 69115 Heidelberg,
Germany. If you have any concerns regarding our products, please
contact ProductSafety@springernature.com

Printed and bound by CPI Group (UK) Ltd, Croydon, CR0 4YY
30/09/2026
02225456-0010